MASTER GUIDE TO
Electronics Circuits

Harry L. Helms

Prentice Hall, ENGLEWOOD CLIFFS, N.J. 07632

Library of Congress Cataloging-in-Publication Data
Helms, Harry L.

   Master guide to electronics circuits.

   Includes index.
   1. Electronic circuits—Handbooks, manuals, etc.
I. Title.
TK7867.H445   1988   621.3815'3   87-17568
ISBN 0-13-559790-0

Cover design: 20/20 Services, Inc.
Manufacturing buyer: Paula Benevento

 © 1988 by Prentice Hall
A Division of Simon & Schuster
Englewood Cliffs, New Jersey 07632

Printed in the United States of America

10 9 8 7 6 5 4 3 2 1

ISBN 0-13-559790-0   025

Prentice-Hall International (UK) Limited, *London*
Prentice-Hall of Australia Pty. Limited, *Sydney*
Prentice-Hall Canada Inc., *Toronto*
Prentice-Hall Hispanoamericana, S.A., *Mexico*
Prentice-Hall of India Private Limited, *New Delhi*
Prentice-Hall of Japan, Inc., *Tokyo*
Simon & Schuster Asia Pte. Ltd., *Singapore*
Editora Prentice-Hall do Brasil, Ltda., *Rio de Janeiro*

This book is dedicated to Cazzie, Mabel, N.O.,

Duckworth—oh yes, and Tina too.

Contents

Preface

Master Guide to Electronics Circuits is a compilation of the latest circuit designs and applications, which have appeared in recent electronics magazines, applications notes, and databooks. As you can see, these circuits cover the entire range of contemporary electronics technology. Each circuit includes part numbers or values for all significant components, an identifying title, a brief description, and a citation of the original source for the circuit. Each circuit diagram has been reproduced directly from the original source without alteration, and this accounts for the differences in schematic style between the various circuits.

The citation style used for circuits taken from magazines is to cite the month and year of publication along with the pages on which the original article featuring the circuit appeared. For circuits taken from an applications note, the company issuing the note and the identifying number of the applications note are given. For circuits taken from a databook, the company name, databook title, and page on which the circuit appeared in the databook are given.

Back issues of the magazines cited should be available at most engineering or larger public libraries. If not, back issues of a particular magazine may be available; a note to the appropriate magazine, along with a self-addressed stamped reply envelope, will let you know if the issue is available and its cost. Applications notes and databooks are available from the issuing company. In some cases these are free (particularly if a request is made on company or professional letterhead), while some companies may charge a nominal fee to cover their printing and shipping costs. Some companies also offer subscription services so you can receive their latest technical data as it is issued. Subscriptions to such services plus the magazines cited in the book make wise investments if you are actively involved in electronics. Addresses for inquiries are given in the acknowledgments which follow.

I hope this compilation saves you hours of searching through literature to locate a specific circuit application.

Harry L. Helms

Acknowledgments

This book would not have been possible without the cooperation and participation of the following companies and publications. *All circuits appearing in this book have been reproduced by the express written permission of each company below, and any further reproduction of any circuit without the permission of the company concerned is prohibited.* The companies below reserve all their rights in regard to the circuits in this book, and any inquiries regarding further reproduction or use of the circuits should go to the company concerned.

We are extremely grateful to the following companies and publications:

Burr-Brown, Inc., International Airport Industrial Park, Tucson, AZ 85734

CQ Magazine, 76 North Broadway, Hicksville, NY 11801

Exar, Inc., 750 Palomar Avenue, P. O. Box 3575, Sunnyvale, CA 94088-3575

GE/Intersil, 10600 Ridgeview Court, Cupertino, CA 95014

GE/RCA Solid State, Route 202, Somerville, NJ, 08876

Gould/AMI Semiconductors, 3800 Homestead Avenue, Santa Clara, CA 95051

ham radio Magazine, Greenville, NH 03048-0498

Linear Technology Corp., 1630 McCarthy Boulevard, Milpitas, CA 95035-7487

QST Magazine, 225 Main Street, Newington, CT 06111

Motorola Semiconductor, P. O. Box 20912, Phoenix, AZ 85036-0924

73 Magazine, WGE Center, Peterborough, NH 03458-1194

Signetics Corp., P. O. Box 3409, 811 East Arques Avenue, Sunnyvale, CA 94088-3409

Abbreviations

The following is a list of abbreviations used in the text of this book and accompanying circuit diagrams. Since the circuit diagrams have been compiled from several different sources, you may see some abbreviations with inconsistent style. For example, "bits per second" may be abbreviated "bps" in one diagram but "B.P.S." in another. The meaning in the both cases will be the same, however.

A	*ampere*
AC	*alternating current*
AC/DC	*AC or DC operation*
A/D	*analog-to-digital*
ADC	*analog-to-digital converter*
AF	*audio frequency*
AFC	*automatic frequency control*
AFSK	*audio frequency shift-keying*
AFT	*automatic fine tuning*
AGC	*automatic gain control*
ALC	*automatic level control*
AM	*amplitude modulation*
ASCII	*American Standard Code for Information Interchange*

AVC	*automatic volume control*
BCD	*binary-coded decimal*
BFO	*beat-frequency oscillator*
BiMOS	*device using both bipolar and MOS transistors*
bps or b/s	*bits per second*
C	*capacitor or capacitance*
CdS	*cadmium sulfide*
cm	*centimeter*
CMOS	*complimentary MOS*
COR	*carrier-operated relay*
CT	*center-tapped or continuous turns*
CW	*continuous wave (i.e., Morse code transmissions)*
D/A	*digital-to-analog*
DAC	*digital-to-analog converter*
dB	*decibel(s)*
dBm	*decibel(s) above 1 mW*
DC	*direct current*
DIP	*dual in-line package*
DPDT	*double-pole, double-throw switch*

DPST	double-pole, single-throw switch		MOS	metal oxide semiconductor
DSB	double sideband		mm	millimeter
DTMF	dual tone multifrequency		ms	millisecond
DVM	digital voltmeter		MSB	most significant bit
ECL	emitter-coupled logic		mH	milliHenry
EMF	electromotive force		mV	millivolt
EMI	electromagnetic interference		mW	milliwatt
EPROM	erasable PROM		μA	microampere
F	farad		μF	microfarad
FET	field-effect transistor		μH	microHenry
FM	frequency modulation		μs	microsecond
FSK	frequency-shift keying		μV	microvolt
GHz	gigahertz		μW	microwatt
H	Henry		NRZ	nonreturn to zero
Hz	Hertz		ns	nanosecond
IC	integrated circuit		PC	printed circuit, or personal computer, or pulse-coded
IF	intermediate frequency		PCM	pulse-coded modulation
IMD	intermodulation distortion		PEP	peak envelope power
I/O	input/output		pF	picofarad
JFET	junction FET		PIN	special type of diode
K	kilohm or kilobyte		PIV	peak-inverse voltage
kb	kilobyte or kilobaud		PLL	phase-locked loop
kHz	kiloHertz		PM	phase modulation
kw	kilowatt		PMOS	P-channel MOS
LCD	liquid crystal display		ppm	parts per million
LED	light-emitting diode		PSK	phase-shift keying
LF	low frequency		p-t-p	peak-to-peak
LSB	lower sideband or least significant bit		PROM	programmable read-only memory
LSI	large scale integration		PTT	push to talk
m	milli		PWM	pulse-width modulation
M	mega or megohm		Q	quality factor
MHz	megaHertz		QRP	low power transmitting

RF	radio frequency		TTL	transistor-transistor logic
RFI	radio frequency interference		TVI	television interference
RIAA	Recording Industry Association of America		UART	universal asynchronous receiver-transmitter
RMS	root-mean square		UHF	ultra high frequency
ROM	read-only memory		V	volt
RTTY	radioteleprinter		VCA	voltage-controlled amplifier
SCA	subsidiary-communications authorization		VCO	voltage-controlled oscillator
			VFO	variable frequency oscillator
SCR	silicon-controlled rectifier		VHF	very high frequency
S-meter	signal strength meter		VLF	very low frequency
SNR	signal-to-noise ratio		VMOS	vertical MOS
SPDT	single-pole, double-throw switch		VOM	volt-ohm meter
			VOX	voice operated
SPST	single-pole, single-throw switch		VU	volume unit
SSB	single sideband		VXO	variable crystal-oscillator
SSTV	slow-scan television		W	watt
SWR	standing wave ratio		WPM	words per minute
TMOS	enhanced special power FET			

1

Amplifiers, Audio

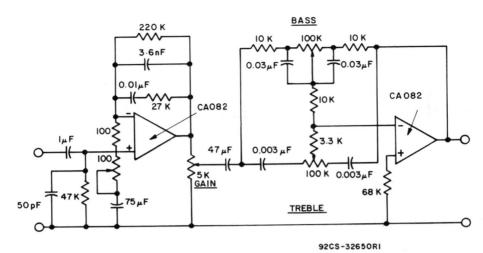

92CS-32650RI

Audio Preamplifier. Both sections of a dual CAO82 BiMOS operational amplifier are used to boost or cut the treble and bass portions of an audio signal before additional amplification. Supply voltage of CAO82 can be ± 18V.—"CAO81, CAO82, CAO84 BiMOS Operational Amplifiers," File #1238, GE/RCA Solid State.

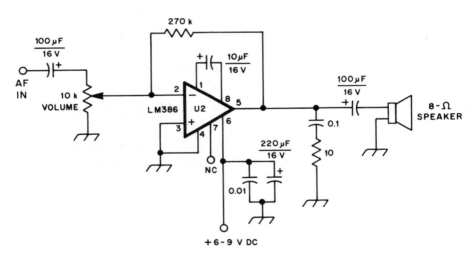

Audio Amplifier. LM386 audio amplifier IC is used to amplify audio input signal to a level sufficient to drive an 8-ohm speaker.—B. Williams, "The SIMPLEceiver," *QST*, September 1986, pp. 34–39.

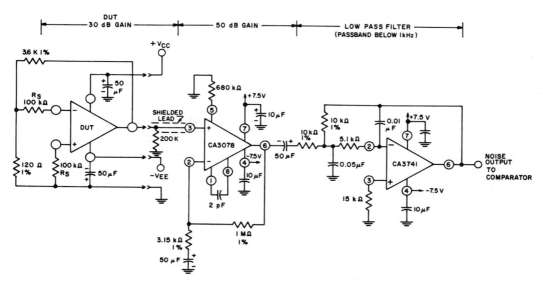

High-Gain Amplifier and Filter. Offers fixed gain of 80 dB with a 12-dB octave roll-off above 1 kHz. Designed to be part of a test procedure involving burst noise. The gain function is arbitrarily distributed between the DUT ("device under test") and post-amplifier at 30 dB and 50 dB respectively.— "Measure of Burst (Popcorn) Noise in Linear Integrated Circuits," ICAN-6732, GE/RCA Solid State.

Audio Amplifier. Configuration can provide a gain of approximately 20; all capacitors should have a working voltage of at least 25V. The 25-K potentiometer controls the output of this circuit.—D. Malley. "Split That Audio," 73 *Magazine,* December 1986, p. 38.

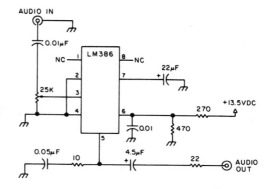

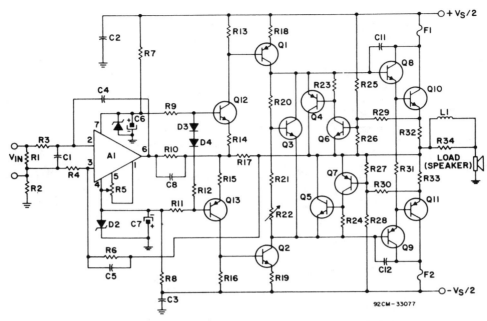

92CM-33077

DC-Coupled 100-Watt Audio Amplifier. CA3100 IC is used as the input stage while the power stage uses discrete transistors; in effect, the circuit is two cascaded gain-blocks with a common feedback loop. The discrete gain-block has its own local feedback provided by R17, R10, and C8. The input stage of the discrete section is a common base stage (Q12, Q13), which serves as a voltage translator and is oper-

ated at class A. The next stage (Q1, Q2) is also operated at class A and its main function is voltage amplification. Accompanying table gives parts values (*facing page*).—"One Hundred Watt True Complementary Symmetry Audio Amplifier Using BD750 and BD751 Silicon Transistors," AN-6904, GE/RCA Solid State.

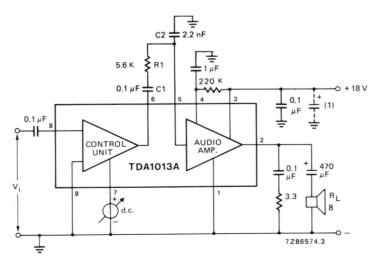

7Z86574.3

Voltage-Controlled 4-W Audio Amplifier— TDA1013A audio amplifier device can produce up to 4-W output depending on a DC voltage of 3.5 – 8V. The control characteristic is logarithmic with a range of more than 80 dB.—"TDA1013A 4W Audio Amplifier with DC Volume Control," 1985 Linear LSI Data and Applications Manual, 6–91, Signetics Corporation.

Components	4 Ohms	8 Ohms
R1 (note 1)	10 K	10 K
R2	1	1
R3	1 K	1 K
R4	220	220
R5 (note 2)	Pot, 10 K	Pot, 10 K
R6	8.2 K	8.2 K
R7	1 K, 1W	1.8 K, 1W
R8	1 K, 1W	1.8 K, 1W
R9	1.8 K	1.8 K
R10	2.2 K	2.2 K
R11	1.8 K	1.8 K
R12	220	220
R13	4.7 K	1.8 K
R14	820	820
R15	820	820
R16	4.7 K	1.8 K
R17	39 K	39 K
R18	47	47
R19	47	47
R20	390	1 K
R21	56	56
R22 (note 3)	Pot, 1 K	Pot, 1 K
R23	100	100
R24	100	100
R25	3.9 K, 1W	8.2 K, 1W
R26	50	68
R27	50	68
R28	3.9 K, 1W	8.2 K, 1W
R29	180	470
R30	180	470
R31 (note 7)	100	100
R32	0.27, 7W	0.68, 7W
R33	0.27, 7W	0.68, 7W
R34	4.7, 1W	10, 1W

	4 Ohms	8 Ohms
C1	100pF	100pF
C2	$0.47\mu F$, 50V	$0.47\mu F$, 50V
C3	$0.47\mu F$, 50V	$0.47\mu F$, 50V
C4	12pF	12pF
C5	100pF	100pF
C6	$22\mu F$, 25V	$22\mu F$, 25V
C7	$22\mu F$, 25V	$22\mu F$, 25V
C8	10nF	10nF
C11 (note 7)	3.9nF	3.9nF
C12 (note 7)	3.9nF	3.9nF
D1	Zener, 15V	Zener 15V
D2	Zener, 15V	Zener 15V
D3	1N4148	1N4148
D4	1N4148	1N4148
Q1 (note 4)	RCA1A10	RCA1A10
Q2 (note 4)	RCA1A11	RCA1A11
Q3 (note 5)	RCA1A18	RCA1A18
Q4	RCP700A	RCP700A
Q5	RCP701A	RCP701A
Q6	RCA1A18	RCA1A18
Q7	RCA1A19	RCA1A19
Q8 (note 6)	RCA1C03	2N6474
Q9 (note 6)	RCA1C04	2N6476
Q10 (note 6)	BD751B	BD751C
Q11 (note 6)	BD750B	BD750C
Q12 (note 4)	RCA1A11	RCA1A11
Q13 (note 4)	RCA1A10	RCA1A10
A1	CA3100	CA3100
F1	4A	3A
F2	4A	3A
L1	$2\mu H$	$4\mu H$
V_S	78V	104V

Notes for Parts List

1. All resistors are non-inductive.
2. Adjust for an output of zero volts with zero volts at the input.
3. Adjust for a quiescent current of 200 mA.
4. Mount each device on heatsink of 30 cm^2 minimum area.
5. Mount on same heatsink as driver and output devices Q_8, Q_9, Q_{10} and Q_{11}.
6. Provide heatsinking as described in text.
7. These components cannot be found on the components layout of Fig. A-1. They are to be mounted directly on the driver-device sockets that are fixed on the heatsink.

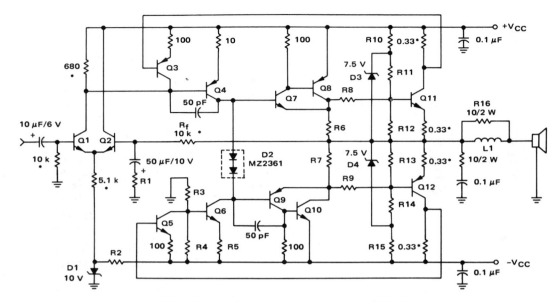

NOTE 1: All of the resistors with the values shown are ±10% tolerance,
          except where * indicates ±5%.
     2: L1 is #20 wire close-wound for the full length of resistor, R16.

High Power Audio Amplifier.   Circuit output can be varied from 35–100 W by using the semiconductors and resistors specified in the accompanying tables. Distortion at 100 W into an 8-ohm load is less than 0.2%. Full short-circuit protection is included. Input sensitivity is 1V for full output, with a frequency response of less than 3-dB rolloff from 10 Hz to 100 kHz (referenced to 1 kHz).—"High Power Audio Amplifiers with Short Circuit Protection," AN-485, Motorola, Inc.

Output Power (Watts-rms)	Load Impedance (Ohms)	R1 ±5%	R2 ±10%	R3 ±5%	R4 ±5%	R5 ±5%	R6, R7 ±5%	R8, R9 ±10%	R10, R15 ±5%	R11, R14 ±5%	R12, R13 ±5%	V_CC
35	4	820	2.7 k	18 k	1.2 k	120	0.39	390	2.7 k	1.5 k	470	±21 V
	8	560	3.9 k	22 k	1.2 k	180	0.47*	240	3.0 k	1.2 k	470	±27 V
50	4	680	3.3 k	22 k	1.2 k	100	0.33	360	3.3 k	1.5 k	470	±25 V
	8	470	4.7 k	27 k	1.2 k	150	0.43*	270	3.9 k	1.2 k	470	±32 V
60	4	620	3.9 k	22 k	1.2 k	120	0.33	430	3.9 k	1.5 k	470	±27 V
	8	430	5.6 k	33 k	1.2 k	120	0.39	300	4.7 k	1.2 k	470	±36 V
75	4	560	4.7 k	27 k	1.2 k	91	0.33	620	5.6 k	1.8 k	470	±30 V
	8	390	6.8 k	33 k	1.2 k	150	0.39	390	6.8 k	1.5 k	470	±40 V
100	4	470	5.6 k	33 k	1.2 k	68	0.39	1.0 k	8.2 k	2.2 k	470	±34 V
	8	330	8.2 k	39 k	1.2 k	100	0.39	510	9.1 k	1.8 k	470	±45 V

NOTE: All of the above resistor values are in ohms and are 1/2 W except for R6 and R7.

*R6 and R7 are 5 W resistors except where * indicates 2 W.

Output Power (Watts-rms)	Load Impedance (Ohms)	Output Transistors		Driver Transistors		Pre-Driver Transistors		Differential Amplifier Transistors
		NPN (Q10)	PNP (Q8)	NPN (Q7)	PNP (Q9)	NPN (Q6)	PNP (Q4)	(Q1 & Q2)
35	4	2N5877	2N5875	MPSU05	MPSU55	MPSA05	MPSA55	MD8001
	8	MJE2801T	MJE2901T	MPSU05	MPSU55	MPSA06	MPSA56	MD8001
50	4	2N5302	2N4399	MPSU05	MPSU55	MPSA06	MPSA56	MD8001
	8	2N5878	2N5876	MPSU06	MPSU56	MPSA06	MPSA56	MD8002
60	4	2N5302	2N4399	MPSU06	MPSU56	MPSA06	MPSA56	MD8001
	8	2N5878	2N5876	MPSU06	MPSU56	MPSA06	MPSA56	MD8002
75	4	MJ802	MJ4502	MPSU06	MPSU56	MPSA06	MPSA56	MD8001
	8	MJ802	MJ4502	MM3007	2N5679	MM3007	MM4007	MD8003
100	4	MJ802	MJ4502	MPSU06	MPSU56	MPSU06	MPSU56	MD8002
	8	MJ802	MJ4502	MM3007	2N5679	MM3007	MM4007	MD8003

The following semiconductors are used at all of the power levels:

Q11 — MPSL01   D1 — 1N5240A or 1N968A (See Note 1)
Q5 — MPSA20   D2 — MZ2361
Q12 — MPSL51   D3 & D4 — 1N5236B (See Note 1)
Q3 — MPSA70

NOTE 1:   For a low-cost zener diode, an emitter-base junction of a silicon transistor can be substituted. A transistor similar to the MPS6512 can be used for the 7.5 V zener.

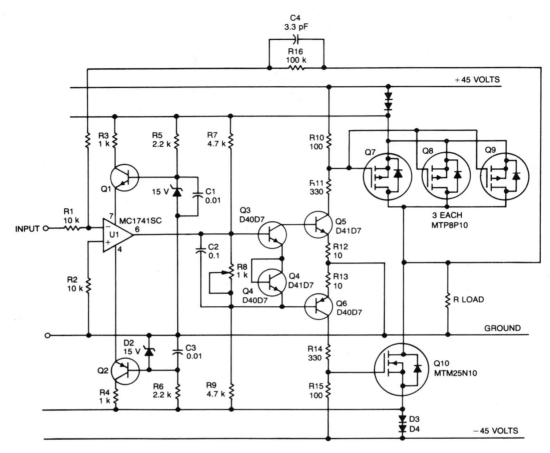

50-W Audio-Power Amplifier. TMOS FET devices operate in a complementary common-source configuration, are biased to cutoff, and turn on very quickly when a signal is applied. Switching speed is very fast in order to obtain very low harmonic and phase distortion. Input source impedance must be less than 100 ohms.—J. Grimes, "50 W Audio Power Amplifier," TMOS Power FET Design Ideas, p. 1, Motorola, Inc.

Audio Amplifier and ALC/AGC Section. Designed for use in a SSB transceiver, circuit offers 400 mW of audio output along with AGC action on receive and ALC on transmit. AGC voltage is sampled from the output of the LM386, detected, and fed to the MPF102. The RC time constant is switched to provide slow release-time during receive and fast release-time during transmit. The 2N3906 provides additional amplification of the control signal, sets the AGC threshold for the LM386, and drives meter M1, which serves as an S-meter on receive and an ALC indicator on transmit. M1 should have a 200-μA movement.—R. Littlefield, "A Compact 75-Meter Monoband Transceiver," *ham radio*, November 1985, pp. 13–27.

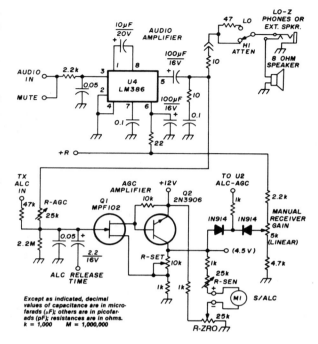

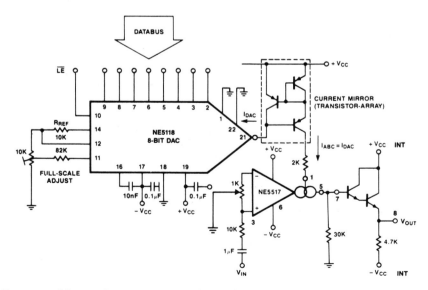

Digitally Programmable Amplifier. The NE5118 8-bit DAC has a current output to control the gain of one half of a NE5517 dual transconductance op amp. The maximum output current of the NE5118 is approximately 1 mA. Circuit is fully micropro- cessor compatible.—"NE5517/5517A Dual Operational Transconductance Amplifier," 1985 Linear LSI Data and Applications Manual, pp. 6–115, Signetics Corporation.

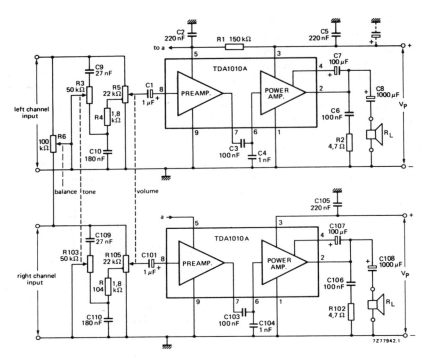

6-W Stereo Audio Amplifier. Two TDA1010A audio-amplifier ICs provide 6W/channel into 4-ohm or 8-ohm speaker loads. Supply voltage can range from 6 – 24 V. Circuit was originally designed for use in auto sound systems—"TDA1010A 6 W Audio Amplifier," 1985 Linear Data and Applications Manual, Volume II, pp. 6–19, Signetics Corporation.

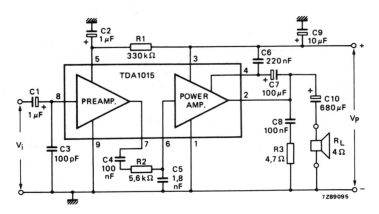

1–4-W Amplifier. TDA1015 audio-amplifier IC provides 1–4-W output depending on a supply voltage of 3.6–18V. Frequency response can range from 60 Hz to 15 kHz.—"TDA1015 1 to 4 W Audio Amplifier," 1985 Linear Data and Applications Manual, Volume II, pp. 6–29, Signetics Corporation.

2
Amplifiers, RF Power

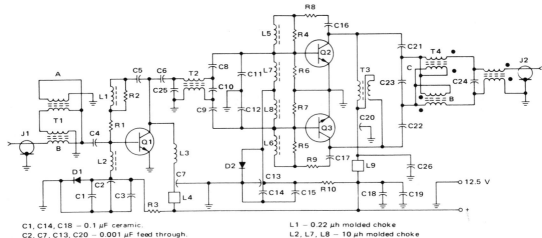

C1, C14, C18 — 0.1 μF ceramic.
C2, C7, C13, C20 — 0.001 μF feed through.
C3 — 100 μF/3V.
C4, C6 — 0.033 μF mylar
C5 — 0.0047 μF mylar.
C8, C9 — 0.015 and 0.033 μF mylars in parallel.
C10 — 470 pF mica.
C11, C12 — 560 pF mica.
C15 — 1000 μF/3 V
C16, C17 — 0.015 μF mylar
C19 — 10 pF 15 V
C21, C22 — two 0.068 μF mylars in parallel.
C23 — 330 pF mica
C24 — 39 pF mica
C25 — 680 pF mica
C26 — .01 μF ceramic

R1, R6, R7 — 10 Ω, 1/2 W carbon.
R2 — 51 Ω, 1/2 W carbon
R3 — 240 Ω, 1 wire W
R4, R5 — 18 Ω, 1 W carbon
R8, R9 — 27 Ω, 2 W carbon
R10 — 33 Ω, 6 W wire W

L1 — 0.22 μh molded choke
L2, L7, L8 — 10 μh molded choke
L5, L6 — 0.15 μh
L3 — 25 t, #26 wire, wound on a 100 Ω, 2 W resistor. (1.0 μh)
L4, L9 — 3 ferrite beads each.

T1 — 2 twisted pairs of #26 wire, 8 twists per inch. A = 4 turns,
     B = 8 turns. Core· ·Stackpole 57·9322·11, Indiana General
     F 627·8Q1 or equivalent.
T2 — 2 twisted pairs of #24 wire, 8 twists per inch, 6 turns.
     (Core as above.)
T3 — 2 twisted pairs of #20 wire, 6 twists per inch, 4 turns.
     (Core as above.)
T4 — A and B = 2 twisted pairs of #24 wire, 8 twists per inch.
     5 turns each. C · 1 twisted pair of #24 wire, 8 turns.
     Core · ·Stackpole 57·9074·11, Indiana General F 624·19Q1
     or equivalent.

Q1 — 2N6367

Q2, Q3 — **MRF460**

D1 — 1N4001
D2 · 1N4997   J1, J2 — BNC connectors

80-W Linear Amplifier for 3–30 MHz. Two MRF463 transistors are used in a push-pull configuration to produce 80-W PEP output. IMD is −30 dB or better. Drive requirements range from 5 W at 30 MHz to 2–3 W at lower frequencies. Original applications note includes details on construction as well as PC-board artwork.—"Broadband Linear Power Amplifiers Using Push-Pull Transistors," AN-593, Motorola, Inc.

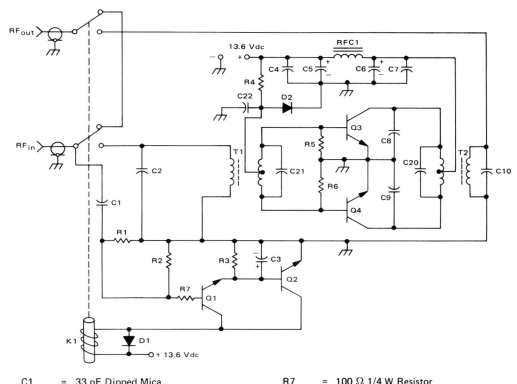

C1	=	33 pF Dipped Mica	R7	=	100 Ω 1/4 W Resistor
C2	=	18 pF Dipped Mica	RFC1	=	9 Ferroxcube Beads on #18 AWG Wire
C3	=	10 μF 35 Vdc for AM operation,	D1	=	1N4001
		100 μF 35 Vdc for SSB operation.	D2	=	1N4997
C4	=	.1 μF Erie	Q1, Q2	=	2N4401
C5	=	10 μF 35 Vdc Electrolytic	Q3, 4	=	MRF454
C6	=	1 μF Tantalum	T1, T2	=	16:1 Transformers
C7	=	.001 μF Erie Disc	C20	=	910 pF Dipped Mica
C8, 9	=	330 pF Dipped Mica	C21	=	1100 pF Dipped Mica
R1	=	100 kΩ 1/4 W Resistor	C10	=	24 pF Dipped Mica
R2, 3	=	10 kΩ 1/4 W Resistor	C22	=	500 μF 3 Vdc Electrolytic
R4	=	33 Ω 5 W Wire Wound Resistor	K1	=	Potter & Brumfield
R5, 6	=	10 Ω 1/2 W Resistor			KT11A 12 Vdc Relay or Equivalent

140-W PEP Linear Amplifier for 2–30 MHz. Two MRF454 power transistors provide 140-W PEP output for an input of 5 W. Operation across 2–30 MHz exhibits a relatively flat response. Transformers T1 and T2 achieve a 16:1 impedance ratio by being wound at a 4:1 ratio. T1 is made from FairRite #77 ferrite beads, 0.375-in. outer diameter by 0.187/ 0.200-in. inner diameter by 0.44in. long. T2 is made from #57-3238-7D ferrite sleeves. Original applications note includes construction details and PC board layout art.—"140W (PEP) Amateur Radio Linear Amplifier 2–30 MHz," EB-63, Motorola, Inc.

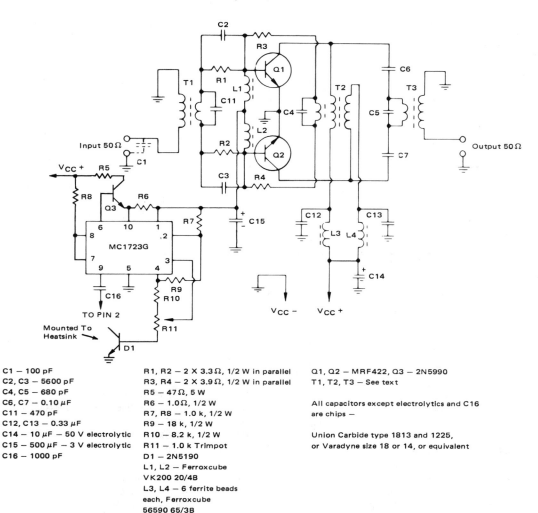

C1 — 100 pF	R1, R2 — 2 X 3.3 Ω, 1/2 W in parallel	Q1, Q2 — MRF422, Q3 — 2N5990
C2, C3 — 5600 pF	R3, R4 — 2 X 3.9 Ω, 1/2 W in parallel	T1, T2, T3 — See text
C4, C5 — 680 pF	R5 — 47 Ω, 5 W	
C6, C7 — 0.10 μF	R6 — 1.0 Ω, 1/2 W	All capacitors except electrolytics and C16
C11 — 470 pF	R7, R8 — 1.0 k, 1/2 W	are chips —
C12, C13 — 0.33 μF	R9 — 18 k, 1/2 W	
C14 — 10 μF — 50 V electrolytic	R10 — 8.2 k, 1/2 W	Union Carbide type 1813 and 1225,
C15 — 500 μF — 3 V electrolytic	R11 — 1.0 k Trimpot	or Varadyne size 18 or 14, or equivalent
C16 — 1000 pF	D1 — 2N5190	
	L1, L2 — Ferroxcube	
	VK200 20/4B	
	L3, L4 — 6 ferrite beads	
	each, Ferroxcube	
	56590 65/3B	

2–30-MHz, 300-W PEP Linear Amplifier.
MRF422 high power RF transistors are used in a
push-pull configuration to provide 300-W PEP in
SSB or CW service. IMD distortion is typically -33
dB or better. T1 is wound on a 57-1845-24B dual-
balun ferrite core. The secondary is made from
⅛-in. copper braid, through which three turns of the
primary (#22 Teflon- insulated wire) are threaded.
T2 is five turns of two twisted pairs of #22 enameled
wire on a 57-9322 toroid. T3 is formed using two
57-3238 ferrite sleeves cemented together; the primary is made of 0.25-in. copper braid through which
three turns of #16 Teflon insulated wire are threaded
as the secondary. Adequate heatsinking of the transistors is vital. Original applications note includes coil
winding details as well as PC board outlines.—"Get
300 Watts PEP Linear Across 2 to 30 MHz From
This Push-Pull Amplifier," EB-27A, Motorola, Inc.

4-MHz RF Amplifier.
Designed for use in a SSB transceiver, MRF138 T-MOSFET transistor operates as class AB and provides 20 dB of gain at 4 MHz for a 28-V supply. Efficiency is approximately 70%, with an output of 30-W PEP into a 50-ohm load. The output is transformed into 50 ohms through T5, a 4:1 balun. T5 is 14 turns of #24 wire bifilar wound on two FT50-43 forms. RFC2 is nine turns of #24 wound on a BF43-801.—R. Littlefield, A Compact 75-Meter Monoband Transceiver, *ham radio*, November 1985, pp. 13–27.

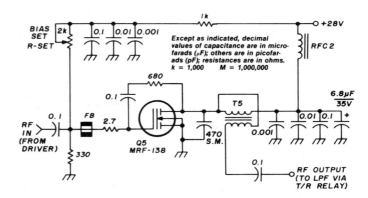

Q5   MRF-138 Motorola "T-MOS" power FET
RFC2   9 turns No. 24 on FB43-801
T5   4:1 balun, 14 turns Bifilar on

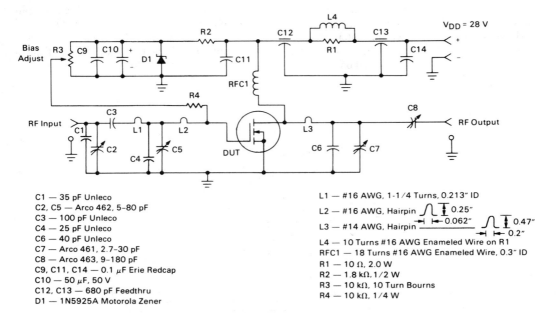

C1 — 35 pF Unleco
C2, C5 — Arco 462, 5–80 pF
C3 — 100 pF Unleco
C4 — 25 pF Unleco
C6 — 40 pF Unleco
C7 — Arco 461, 2.7–30 pF
C8 — Arco 463, 9–180 pF
C9, C11, C14 — 0.1 μF Erie Redcap
C10 — 50 μF, 50 V
C12, C13 — 680 pF Feedthru
D1 — 1N5925A Motorola Zener

L1 — #16 AWG, 1-1/4 Turns, 0.213" ID
L2 — #16 AWG, Hairpin 0.25"
L3 — #14 AWG, Hairpin ⊢ 0.062" 0.47" / 0.2"
L4 — 10 Turns #16 AWG Enameled Wire on R1
RFC1 — 18 Turns #16 AWG Enameled Wire, 0.3" ID
R1 — 10 Ω, 2.0 W
R2 — 1.8 kΩ, 1/2 W
R3 — 10 kΩ, 10 Turn Bourns
R4 — 10 kΩ, 1/4 W

125-W Amplifier for 150 MHz. MRF174 TMOS FET transistor is used in class C amplifier offering a gain of 12 dB and a maximum output power of 125 W. Circuit can operate into a 30:1 SWR mismatch without damage.—"VHF MOS Power Applications," AN-878, Motorola, Inc.

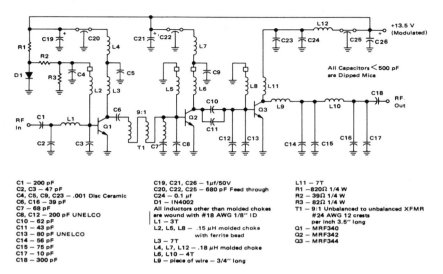

C1 — 200 pF	C19, C21, C26 — 1μf/50V	L11 — 7T
C2, C3 — 47 pF	C20, C22, C25 — 680 pF Feed through	R1 —820Ω 1/4 W
C4, C5, C9, C23 — .001 Disc Ceramic	C24 — 0.1 μf	R2 — 39Ω 1/4 W
C6, C16 — 39 pF	D1 — 1N4002	R3 — 82Ω 1/4 W
C7 — 68 pF	All inductors other than molded chokes	T1 — 9:1 Unbalanced to unbalanced XFMR
C8, C12 — 200 pF UNELCO	are wound with #18 AWG 1/8" ID	#24 AWG 12 crests
C10 — 62 pF	L1 — 3T	per inch 3.5" long
C11 — 43 pF	L2, L5, L8 — .15 μH molded choke	Q1 — MRF340
C13 — 80 pF UNELCO	with ferrite bead	Q2 — MRF342
C14 — 56 pF	L3 — 7T	Q3 — MRF344
C15 — 75 pF	L4, L7, L12 — .18 μH molded choke	
C17 — 10 pF	L6, L10 — 4T	
C18 — 300 pF	L9 — piece of wire — 3/4" long	

15-W AM Amplifier for 118–136-MHz. Covers the 118–136-MHz aircraft band, where AM is the emission mode used. Input signal can be approximately 14–18 mW to produce full output. Circuit uses a MRF340 transistor as a pre-driver, a MRF342 as a driver, and a MRF344 as the final power amplifier. All three transistors are common emitter. Input and output impedances are 50 ohms. Original applications note includes construction data and PC board artwork.—"A 15 Watt AM Aircraft Transmitter Power Amplifier Using Low Cost Plastic Transistors," AN-793, Motorola Inc.

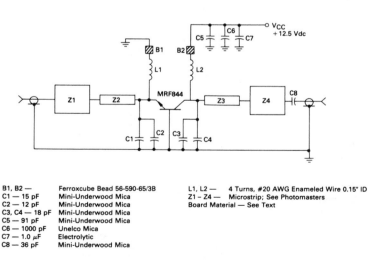

B1, B2 —	Ferroxcube Bead 56-590-65/3B
C1 — 15 pF	Mini-Underwood Mica
C2 — 12 pF	Mini-Underwood Mica
C3, C4 — 18 pF	Mini-Underwood Mica
C5 — 91 pF	Mini-Underwood Mica
C6 — 1000 pF	Unelco Mica
C7 — 1.0 μF	Electrolytic
C8 — 36 pF	Mini-Underwood Mica

L1, L2 —	4 Turns, #20 AWG Enameled Wire 0.15" ID
Z1 - Z4 —	Microstrip; See Photomasters
Board Material — See Text	

30-W Amplifier for 800 MHz. Circuit uses MRF844 transistor in a common base, class C configuration to provide 5-dB gain over a range of 800–870 MHz. Input power should be between 8–10 W. Input and output impedances re 50 ohms. Z1 through Z4 are microstrips on the PC board. Original applications note includes details on construction and component placement alongwith PC board artwork.—"A 30 Watt, 800 MHz Amplifier Design," EB-105, Motorola, Inc.

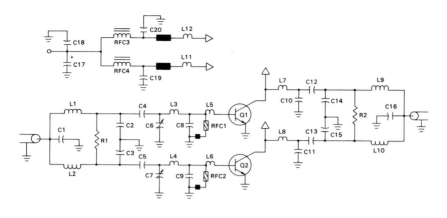

C1, C16 — 25 pF Unelco (J101)
C2, C3 — 15 pF CM04 Mica
C4, C5 — 68 pF Standex
C6, C7 — Arco 404 Variable
C8, C9 — 150 pF Standex
C10, C11 — 56 pF Standex
C12, C13 — 39 pF Standex
C14, C15 — 15 pF Standex
C17 — 100 μF @ 16 V Electrolytic
C18, C19, C20 — 680 pF Allen Bradley Feedthru

L1, L2 — 7 Turns #18, 0.125" ID
L3, L4, L5, L6 — Printed Inductors
L7, L8 — Printed Inductors
L9, L10 — 7 Turns #18 AWG, 0.125 ID
L11, L12 — 4 Turns #18 AWG, 0.250 ID w/Bead
Q1, Q2 — MRF264
RFC1, RFC2 — 0.15 μH Molded Choke w/Bead,
    Ferroxcube 56-590 65/3B
RFC3, RFC4 — 4 Ferrite Beads each on #18 AWG
R1 — 100 Ω 1/2 W Carbon
R2 — 100 Ω 2.0 W Carbon

60-W Amplifier for 150–175 MHz. Two MRF264 transistors are used in a push-pull, class C configuration capable of supplying 60-W output for a 15-W input. Microstrip techniques are used. Original engineering bulletin includes construction data and PC board artwork—"60 Watt VHF Amplifier Uses Splitting/Combining Techniques," EB-93, Motorola, Inc.

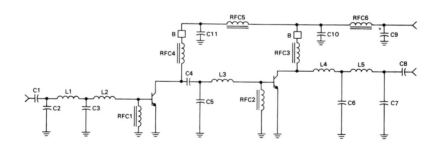

C1 — 220 pF, TDK 100 mil Chip Capacitor
C2 — 43 pF, TDK 100 mil Chip Capacitor
C3 — 150 pF, TDK 100 mil Chip Capacitor
C4 — 15 pF, TDK 100 mil Chip Capacitor
C5 — 63 pF, TDK 100 mil Chip Capacitor
C6 — 27 pF, TDK 100 mil Chip Capacitor
C7 — 22 pF, TDK 100 mil Chip Capacitor
C8 — 100 pF, TDK 100 mil Chip Capacitor
C9 — 1.0 μF Tantalum
C10 — 0.1 μF Erie Redcap, 100 V General Purpose
C11 — 0.05 μF Erie Redcap, 100 V General Purpose

L1–L5 — Printed Inductor
L3 — 5/8" #18 AWG Wire formed into hairpin loop
Q1 — MRF260
Q2 — MRF262
RFC1, RFC2 — 2 Turns #26 Enameled Wire
    through Ferrite Bead Ferroxcube 56-590-65/3B
RFC3 — 0.15 μH Molded Choke
RFC4 — 10 μH Molded Choke
RFC5, RFC6 — VK200-4B
B — Bead, Ferroxcube 56-590-65/3B

10–15-W Amplifier for 160–174 MHz Chip capacitors are used to provide 10–15-W output across 160–174 MHz for 220-mW input. Amplifier operates as class C for FM and CW signals. Original applications note includes construction details and PC board artwork.—"Low Cost VHF Amplifier Has Broadband Performance," EB-90, Motorola, Inc.

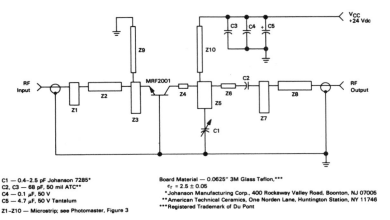

C1 — 0.4–2.5 pF Johanson 7285*
C2, C3 — 68 pF, 50 mil ATC**
C4 — 0.1 µF, 50 V
C5 — 4.7 µF, 50 V Tantalum

Z1–Z10 — Microstrip; see Photomaster, Figure 3

Board Material — 0.0625" 3M Glass Teflon,***
ϵ$_r$ = 2.5 ± 0.05
*Johanson Manufacturing Corp., 400 Rockaway Valley Road, Boonton, NJ 07005
**American Technical Ceramics, One Norden Lane, Huntington Station, NY 11746
***Registered Trademark of Du Pont

1-W Amplifier for 2.3 GHz. Delivers 1-W output at 8-dB gain and is tunable from 2.25–2.35 GHz. Circuit uses a MRF2001 transistor in a common-base configuration at class C operation. Input and output impedances are 50 ohms. Z1 through Z10 are microstrips etched on the PC board. Original engineering bulletin includes PC board artwork and construction details.—"A 1 Watt, 2.3 GHz Amplifier," EB-89, Motorola, Inc.

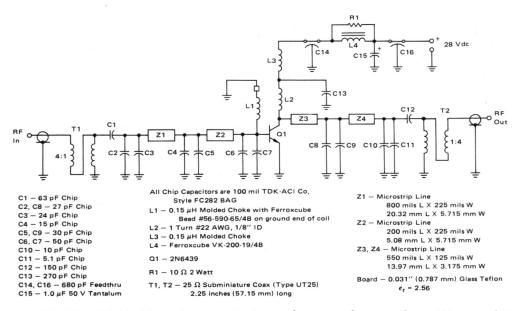

C1 — 63 pF Chip
C2, C8 — 27 pF Chip
C3 — 24 pF Chip
C4 — 15 pF Chip
C5, C9 — 30 pF Chip
C6, C7 — 50 pF Chip
C10 — 10 pF Chip
C11 — 5.1 pF Chip
C12 — 150 pF Chip
C13 — 270 pF Chip
C14, C16 — 680 pF Feedthru
C15 — 1.0 µF 50 V Tantalum

All Chip Capacitors are 100 mil TDK-ACI Co,
Style FC282 BAG
L1 — 0.15 µH Molded Choke with Ferroxcube
Bead #56-590-65/4B on ground end of coil
L2 — 1 Turn #22 AWG, 1/8" ID
L3 — 0.15 µH Molded Choke
L4 — Ferroxcube VK-200-19/4B
Q1 — 2N6439
R1 — 10 Ω 2 Watt
T1, T2 — 25 Ω Subminiature Coax (Type UT25)
2.25 inches (57.15 mm) long

Z1 — Microstrip Line
800 mils L × 225 mils W
20.32 mm L × 5.715 mm W
Z2 — Microstrip Line
200 mils L × 225 mils W
5.08 mm L × 5.715 mm W
Z3, Z4 — Microstrip Line
550 mils L × 125 mils W
13.97 mm L × 3.175 mm W

Board — 0.031" (0.787 mm) Glass Teflon
ϵ$_r$ = 2.56

60-W, 225–400 MHz Amplifier. Input signal of approximately 8.5 W can drive amplifier to 60-W output across 225–400-MHz military communications band. Input and output impedances are 50 ohms. T1 is a 4:1 impedance-ratio coaxial transformer. Two amplifiers may be combined through quadrature coupling to produce a 100-output device. Original applications notes give details on construction and component layout, which can be critical at these frequencies.—"A 60 Watt 225–400 MHz Amplifier," EB-77, Motorola, Inc.

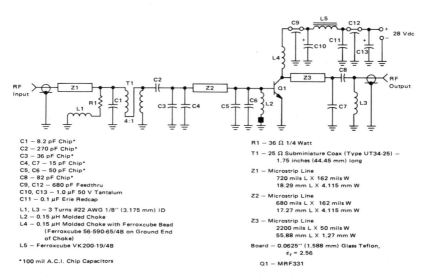

C1 — 8.2 pF Chip*
C2 — 270 pF Chip*
C3 — 36 pF Chip*
C4, C7 — 15 pF Chip*
C5, C6 — 50 pF Chip*
C8 — 82 pF Chip*
C9, C12 — 680 pF Feedthru
C10, C13 — 1.0 μF 50 V Tantalum
C11 — 0.1 μF Erie Redcap

L1, L3 — 3 Turns #22 AWG 1/8″ (3.175 mm) ID
L2 — 0.15 μH Molded Choke
L4 — 0.15 μH Molded Choke with Ferroxcube Bead
    (Ferroxcube 56-590-65/4B on Ground End
    of Choke)
L5 — Ferroxcube VK200-19/4B

*100 mil A.C.I. Chip Capacitors

R1 — 36 Ω 1/4 Watt

T1 — 25 Ω Subminiature Coax (Type UT34-25) —
    1.75 inches (44.45 mm) long

Z1 — Microstrip Line
    720 mils L X 162 mils W
    18.29 mm L X 4.115 mm W

Z2 — Microstrip Line
    680 mils L X 162 mils W
    17.27 mm L X 4.115 mm W

Z3 — Microstrip Line
    2200 mils L X 50 mils W
    55.88 mm L X 1.27 mm W

Board — 0.0625″ (1.588 mm) Glass Teflon,
    ϵ_r = 2.56

Q1 — MRF331

10-W, 225–400-MHz Amplifier. Input power of 300–1200 mW can produce output of up to 10 W from 225–400 MHz, a military communications range. Input and output impedances are 50 ohms. T1 is 4:1 impedance-ratio coaxial transformer. Original applications note contains information on construction and parts placement, both of which can be quite crucial at these frequencies.—"A 10 Watt 225–400 MHz Amplifier," EB-74, Motorola, Inc.

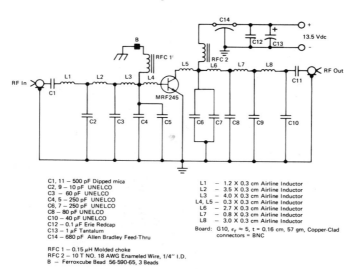

C1, 11 — 500 pF Dipped mica
C2, 9 — 10 pF UNELCO
C3 — 60 pF UNELCO
C4, 5 — 250 pF UNELCO
C6, 7 — 250 pF UNELCO
C8 — 80 pF UNELCO
C10 — 40 pF UNELCO
C12 — 0.1 μF Erie Redcap
C13 — 1 μF Tantalum
C14 — 680 pF Allen Bradley Feed-Thru

RFC 1 — 0.15 μH Molded choke
RFC 2 — 10 T NO. 18 AWG Enameled Wire, 1/4″ I.D.
B — Ferroxcube Bead 56-590-65, 3 Beads

L1 — 1.2 X 0.3 cm Airline Inductor
L2 — 3.5 X 0.3 cm Airline Inductor
L3 — 4.0 X 0.3 cm Airline Inductor
L4, L5 — 0.3 X 0.3 cm Airline Inductor
L6 — 2.7 X 0.3 cm Airline Inductor
L7 — 0.8 X 0.3 cm Airline Inductor
L8 — 3.0 X 0.3 cm Airline Inductor

Board: G10, ϵ_r ≈ 5, t = 0.16 cm, 57 gm, Copper-Clad
    connectors = BNC

80-W Amplifier covering 143–156 MHz. Single MRF245 transistor provides 9.4-dB gain from 143–156 MHz into a 50-ohm load. Eight watts of RF input produces 80-W output. Operation is class C, making it suitable for FM or CW signals. Amplifier can operate into a SWR mismatch of 20:1 without damage. Supply voltage should be 13.5-V DC.—"A Single Device, 80 Watt, 50 Ohm VHF Amplifier," EB-46, Motorola, Inc.

3
Audio Processing Circuits

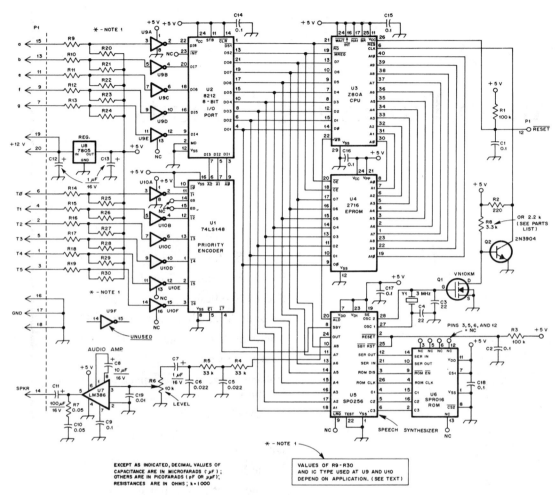

EXCEPT AS INDICATED, DECIMAL VALUES OF
CAPACITANCE ARE IN MICROFARADS (μF) ;
OTHERS ARE IN PICOFARADS (pF OR μμF);
RESISTANCES ARE IN OHMS ; k = 1000

VALUES OF R9 - R30
AND IC TYPE USED AT U9 AND U10
DEPEND ON APPLICATION. (SEE TEXT)

Speech Synthesizer. Designed to announce the frequency to which a receiver or transceiver is tuned, but can be adapted to many other uses. "Heart" of the circuit is a SPO256 voice synthesizer IC and its matching SPRO16 speech ROM; the ROM includes numbers and a few words (the SPO256 and SPRO16 were originally designed for use in "talking clocks"). The Z80 microprocessor "translates" input from the receiver or transceiver frequency display into commands for the SPO256; the Z80 is controlled by the software in the 2716 EPROM (U4). Other applications of this circuit will require new controlling software for the Z80. As each command is received from

the Z80, the appropriate "word" is recalled from the SPRO16 ROM and produces output from the SPO256. The values of R9–R19 and R20–R30 depend upon the voltages required at the outputs of U9 and U10. Values of 120 K for R9–R19 and 47 K for R20–R30 will produce values of + 5 V and zero. A complete kit of parts, including preprogrammed EPROMs, was available at the time of the original article's publication from A & A Engineering, 7970 Orchid Dr., Buena Park, Ca 90620.—J. Langner, "A Talking Frequency Display," QST, April 1985, pp. 14–17.

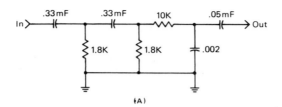

(A)

Passive Microphone-Equalizer. Two-section RC high-pass filter "rolls off" its input frequency below about 1 kHz and a single low-pass section rolls off above 1–2 kHz for a more uniform, "flattened" microphone response. Input is taken from a micro- phone. Output of circuit will normally need some amplification since the passive nature of the circuit introduces some loss.—J. Schultz, "How to Build a Simple Microphone Interface/Test Oscillator Unit," CQ, March 1986, pp. 32–34.

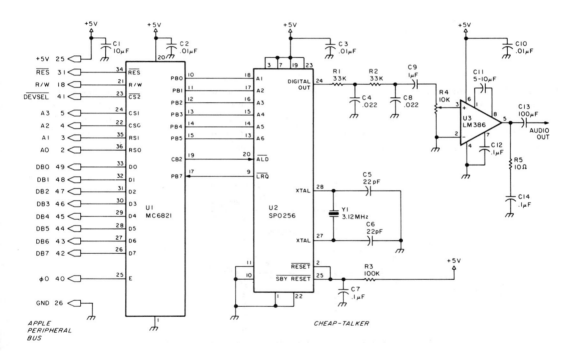

Speech Synthesizer. Based upon SPO256 speech synthesizer IC from General Instruments, this circuit uses an allophone-synthesis method to generate words by stringing together various phoneme codes representing the sounds of English. An MC6821 parallel-interface IC is used to interface the synthesizer to the bus of an Apple microcomputer (the SPO256 can be directly connected to a Centronics-compat- ible parallel interface). Original article includes a BASIC language control-program for the synthesizer along with a table of words and their allophone code equivalents which can be generated by this circuit. Applications information is also provided.—T. Johnson, "Talk is Cheap," 73 Magazine, October 1985, pp. 22–30.

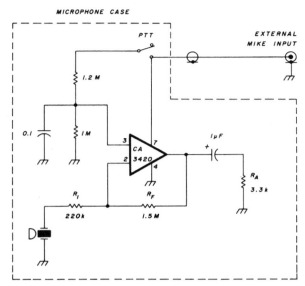

Ceramic Microphone Amplifier. Boosts average 35-mV (into 300 K load) output of typical ceramic microphone to 250 mV. Low power CA3420 requires only 2 V and a current of 350 μA; circuit is powered from the transmitter's microphone input circuitry. Entire circuit fits inside the case of a typical ceramic microphone.—E. Richley and F. Caimi, "Converting Mobile Microphones for Handheld VHF Transceivers," *ham radio*, March 1986, pp. 79–84.

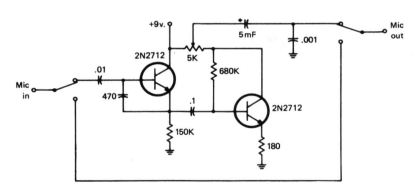

Microphone Preamplifier. Designed for use with high-impedance microphones such as the Astatic D-104, but can be used with most low-impedance types by using a general-purpose interstage transformer with a 1:20 K ratio at the circuit's input.—J. Schultz, "Building ideas for a Consolidated Control Console," *CQ*, July 1986, pp. 11–21.

Cassette Preamplifier. Uses TDA1522 IC, which provides two independent amplifiers with a gain of 90 dB. Total harmonic distortion is 0.05% or less with a channel separation of at least 45 dB. The supply voltage can range from 7.5–23 V. Circuit was designed for car radio/cassette players.— "TDA1522 Cassette Preamplifier," 1985 Linear LSI Data and Applications Manual, pp. 5–51, Signetics Corporation.

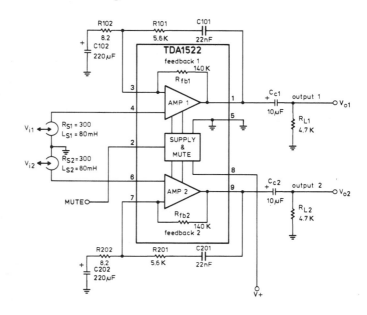

Voltage-to-Frequency Converter for Music Synthesizers An input of 1 V produces a one-octave output with an input range from 0–10 V. Shaded component is a LT1055 op amp. All pin numbered transistors are part of a CA3096 transistor array IC. All asterisked resistors are 1% metal-film resistors.— "LT1055/LT1056 Precision High Speed JFET Input Operational Amplifiers," data sheet, Linear Technology Corporation.

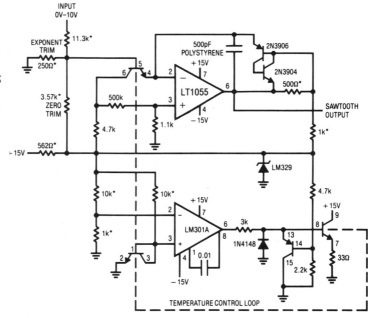

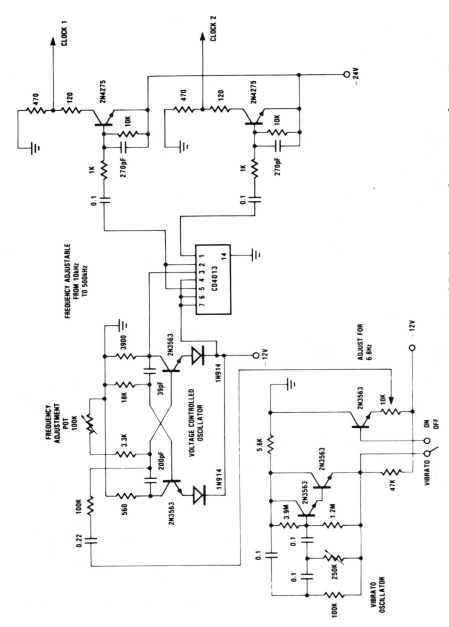

Vibrato Circuit. Used to produce vibrato effects (a small frequency change, usually a few hertz) in music synthesis circuits. Output of circuit is in the form of two clock signals which can be used to frequency- modulate the main synthesis section of a music synthesizer.—"MOS Music," Applications Note, Gould/AMI Semiconductors.

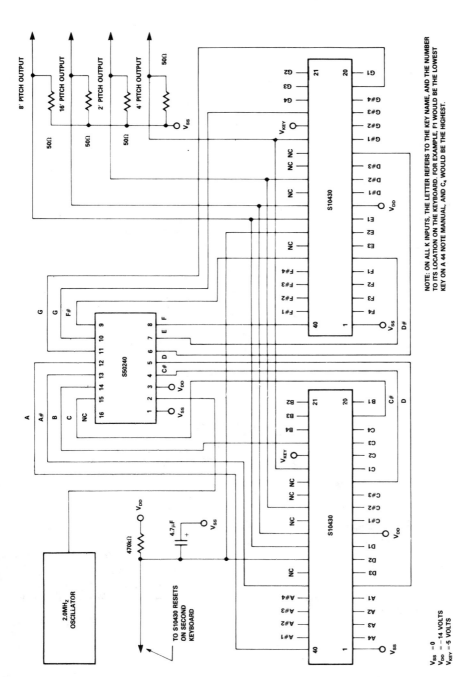

Top Octave Synthesizer with Divider/Keyer. Used in music synthesis circuits. It is a two-tone generator using a S50240 top octave synthesizer and two S10430 divider/keyer circuits to produce a 44-note tone generator. Output of circuit goes to a series of format filters before final amplification and output to speakers.—"MOS Music," Applications Note, Gould/AMI Semiconductors.

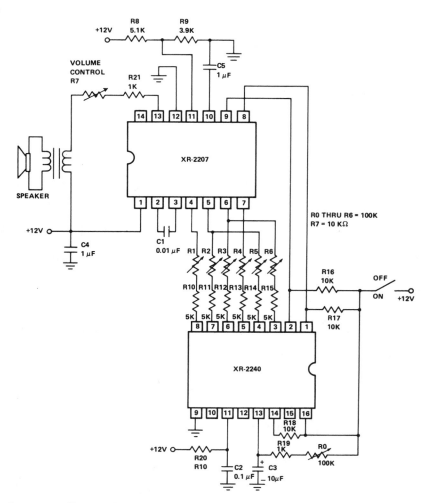

Music Synthesizer. All active components necessary for electronic music synthesis are contained in XR-2207 programmable tone generator and XR-2240 counter/timer device. The XR-2240 produces a pseudorandom binary pulse pattern to drive the XR-2207. The pulse pattern repeats itself at 8-bit (256 count) intervals of the time base period. Thus, the output tone sequence continues for abut 1–2 minutes, depending on the "beat," and then repeats itself. Potentiometers R1 – R6 control the output tones while R10 controls the pulse output of the timer.—"An Electronic Music Synthesizer Using the XR-2207 and the XR-2240," AN-15, Exar Corporation.

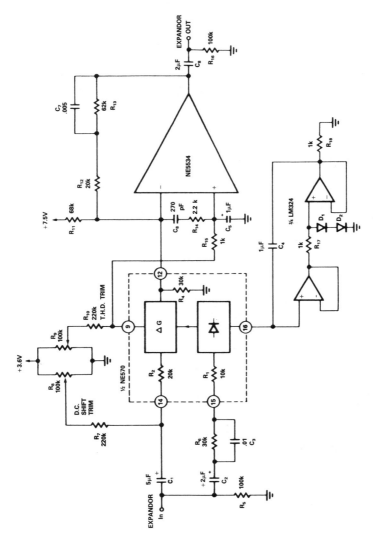

Expandor with De-Emphasis. Restores compandored audio to original form and includes a de-emphasis section to overcome effects of pre-emphasis in the compandor stage. A NE570 compandor IC is used along with an external op amp (NE5534) for a high slew rate. Unity gain level is 0 dBm.—"Applications for Compandors: NE570/571/SA571," AN174, Signetics Corporation.

29

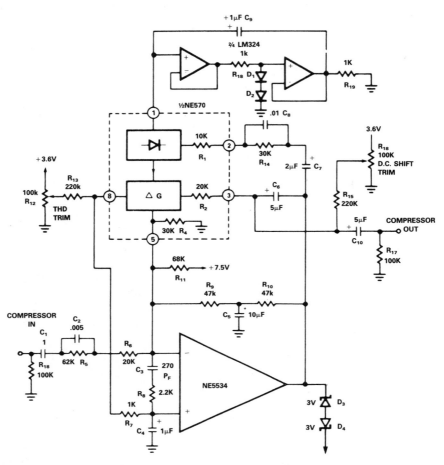

High Fidelity Compandor. Circuit has high gain and wide bandwidth for applications such as noise reduction in tape recorders, transmission systems, bucket brigade delay lines, and digital audio systems. A pre-emphasis stage (consisting of C2, C8, R5, and R14) minimizes "breathing" caused by changes in the background noise level due to changes in the circuit gain. Unity gain level is 0 dBm.—"Applications for Compandors: NE570/571/SA571," AN174, Signetics Corporation.

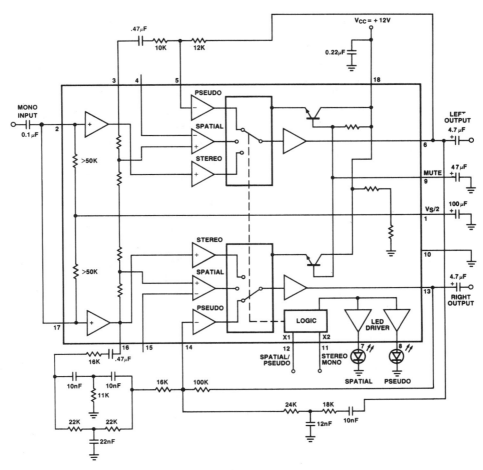

Pseudo-Stereo Processor. Circuit uses a TDA3810 spatial processor to convert a mono-input into a pseudo-stereo output with separate left and right channels. Circuit shown in box is internal diagram of TDA3810.—"TDA3810 Spatial, Stereo, Pseudo-Stereo Processor," 1985 Linear Data and Applications Manual, Volume II, pp. 5–74, Signetics Corporation.

4

Automotive Circuits

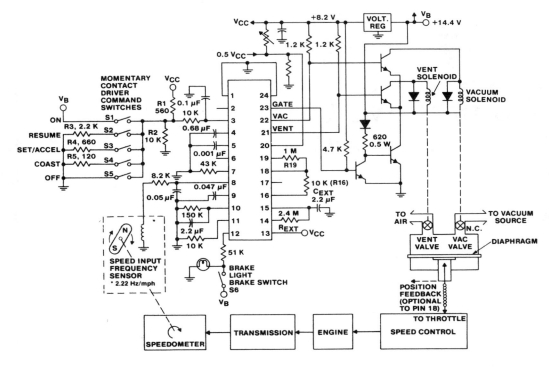

Automotive Cruise Control.   Designed around the CA3228E speed control system, a monolithic IC designed especially for automobile cruise control systems. The CA3228E incorporates a frequency-to-voltage converter stage which accepts changing frequency signals and converts them to their linear representation as DC voltages. The CA3228E also includes a 9-bit analog-to-digital/digital-to-analog conversion stage to interface between the device's control logic and the engine. Other on-chip functions include minimum speed and overspeed detection, brake sensing, and automatic braking. Original applications notes include information on adapting this circuit for control of DC or AC motors, fluid movement rate, and temperature.—"Applications of the CA3228E Speed Control Systems," ICAN-7326, GE/RCA Solid State.

Automobile Voltage Regulator.   The 555 timer IC ▶ is the heart of this circuit. When the timer is off, so that its output is low, the power Darlington transistor pair (MJE1090 or equivalent) is switched off. If the battery voltage becomes less than 14.4 V, the timer turns on and the Darlington pair conducts. —"NE555 and NE556 Applications," AN170, Signetics Corporation.

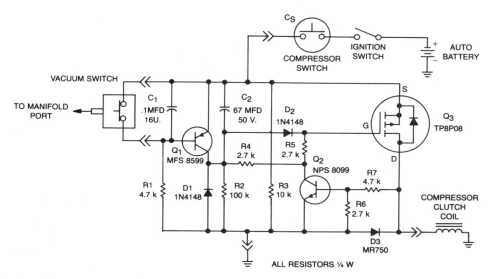

"Smart" Clutch For Auto Air Conditioner. Disables the air conditioner compressor when additional engine power is required for passing, etc. It monitors the engine vacuum at the intake manifold. If vacuum drops to 40% of its normal level, the compressor clutch is disabled, removing the air conditioner load from the engine. After the engine returns to a normal vacuum level, there is a 6-second delay before the compressor clutch is enabled and the air conditioner is reactivated.—S. Sitts, "Automotive Air Conditioner Smart Clutch," *TMOS Power FET Design Ideas*, p. 3, Motorola, Inc.

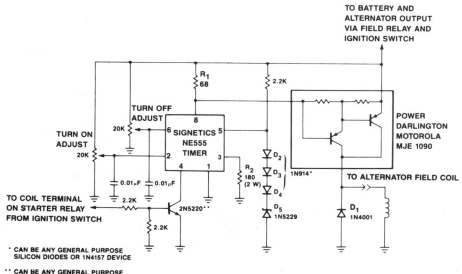

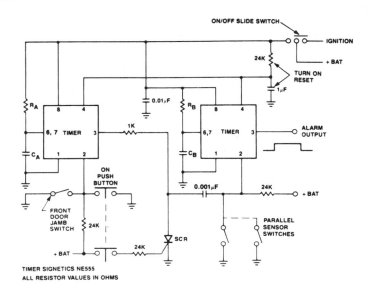

Automobile Burglar Alarm.
Uses a dual 556 timer or two
555 timer devices. Timer A
produces a delay allowing the
driver to disarm the alarm; this
approach eliminates a
vulnerable outside control
switch. The SCR prevents
timer A from triggering timer
B, unless timer B is triggered by
strategically located sensor
switches.—"NE555 and NE556
Applications," AN170,
Signetics Corporation.

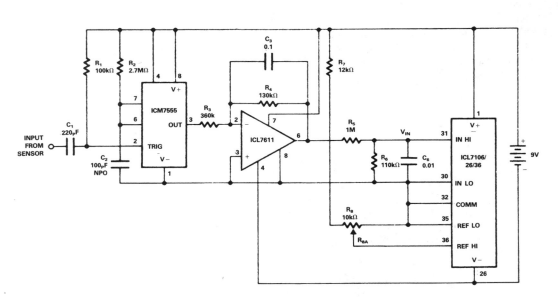

Battery Operated Tachometer.   ICL7106 ADC is
preceded by a frequency-to-voltage conversion stage,
allowing the ADC to read RPM from an input frequency. The ICM7555 timer generates a constant
pulse width waveform at its output; the ICL7611 in-
tegrates the timer pulses. Operating the timer from
the internal-reference voltage of the ADC eliminates
the need for a second reference because of the rail-to-
rail output swing of the timer.—"Games People Play
With Intersil's A/D Converters," A047, GE/Intersil.

5
Battery Circuits

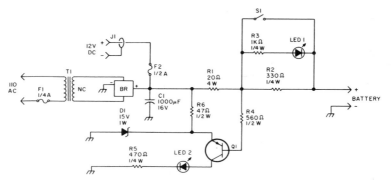

AC/DC "Nicad" Battery Charger. Accepts both 120-V AC and 12-V DC for conversion to recharge 6–12-V nickel-cadmium ("nicad") batteries. The output is 12-V DC at either 180 mA (the "fast charge" rate) or 20 mA (the "slow charge" rate): closing switch S1 selects the higher current output. LED1 lights when unit is in the low current mode while LED2 lights during high current operation. Care must be taken when using the fast charging rate since some nicad cells cannot handle the output current. Q1 is any general purpose PNP silicon transistor. BR is a bridge rectifier rated at 4 A and 50 PIV. T1 is a 120-V primary, 12-V secondary, center-tapped transformer. LED1 is a T-1 ¾ green LED while LED2 is a CQX21 blinking red LED.—D. Olszewski, "Speed-Charge Your PB-21," 73 *Magazine*, May 1986, pp. 32–34.

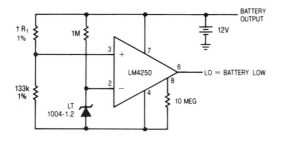

Low-Battery Detector For Lead-Acid Cells. Resistor R1 sets the "trip point" for the circuit, computed as 60.4 K per cell with each cell representing 1.8 V. A low output from the LM4250 indicates a low battery voltage.—"LT1004 Micropower Voltage References," data sheet, Linear Technology Corporation.

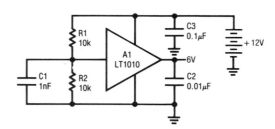

"Battery Splitter" Produces positive and negative supply voltages at low currents from a single battery. Each voltage is equal to one-half of the battery's rated voltage. —"Power Conditioning Techniques for Batteries," Application Note 8, Linear Technology Corporation.

Solar-Cell Battery Charger.
Single solar cell is able to
charge a 9-V battery at about
30 mA per input ampere at
0.4V. U1 is a quad Schmitt
trigger operating as an astable
multivibrator to drive push-pull
TMOS devices Q1 and Q2.
Power for U1 is derived from
the 9-V battery via D4 while
power for Q1/Q2 is supplied by
the solar cell. The multivibrator
frequency is 180 Hz. The CdS
photocell shuts off the oscillator
in darkness to preserve the fail-
safe battery during prolonged
darkness.—P. Smith, "Battery
Charger Operates on a Single
Solar Cell." TMOS *Power FET
Design Ideas*, p. 8, Motorola,
Inc.

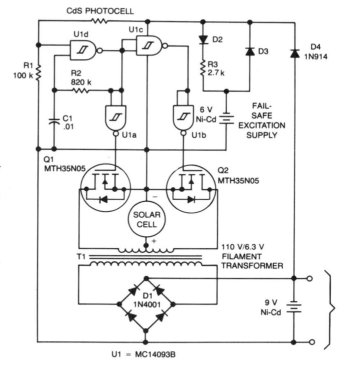

Nicad-Battery Discharge Protection Circuit Circuit detects a drop of 1.25 V in a Nicad cell and turns
the load off before cell-polarity reversal can occur.
Circuit takes advantage of the high input impedance
and low series resistance of a TMOS device to keep
the circuit's power consumption low so that battery
shelf life is not affected. A low current Zener diode
(D1) and resistors R1 and R2 establish a reference
level for transistor Q1. These resistors bias the Zener
to a few microamperes above its "knee"; thus if battery voltage falls more than 1.25 V, Q1 turns off,
turning off Q2, and disconnecting the load.—L.A.
Turner, "NiCad Battery Protection Circuit," TMOS
Power FET Design Ideas, p. 5, Motorola, Inc.

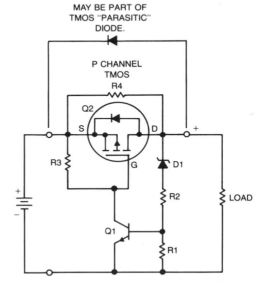

6

Converters, Analog to Digital and Digital to Analog

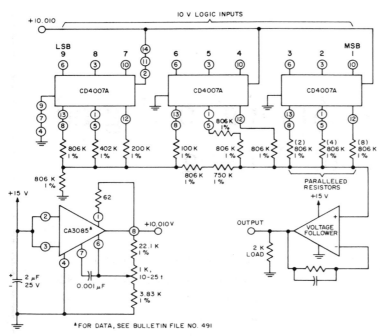

9-Bit Digital-to-Analog Converter. Three 4007 CMOS devices are used in a DAC having output proportional accuracy of ±0.25 LSB. Accuracy can be maintained even if there is a variation of several volts in the supply. The 806-K resistors must be closely matched in value for proper operation.— "Digital to Analog Conversion Using the RCA CD4007A COS/MOS IC," ICAN-6080. GE/RCA Solid State.

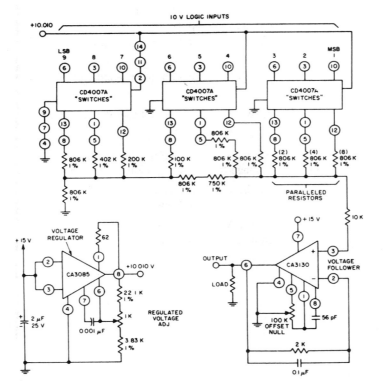

9-Bit Digital-to-Analog Converter. CA3130 BiMOS op amp is connected as a follower and a CA3085 voltage regulator IC is used to provide a constant 10-V level. Three 4007 CMOS "transistor pair" ICs are used as switches in conjunction with the ladder network of precision resistors.—"CA3130 BiMOS Operational Amplifiers," File #817, GE/RCA Solid State.

6-Bit Analog-to-Digital Converter CA3300 6-bit ADC IC offers 15-MHz sampling rate with ±0.5 LSB accuracy. Two units may be operated in a series for 7-bit output while two units in parallel give a 30-MHz sampling rate.—"CA3300D/CA3300J High Reliability Slash (/) Series CMOS Video Speed 6-Bit Flash Analog to Digital Converter," File #1835, GE/RCA Solid State.

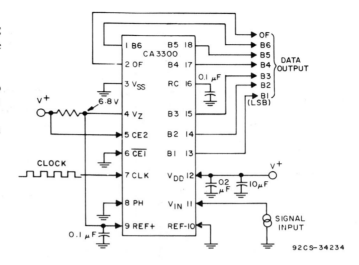

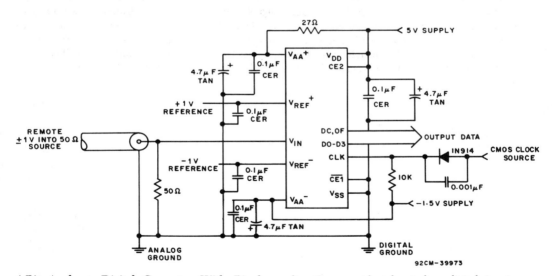

4-Bit Analog-to-Digital Converter With Bipolar Operation. CA3304 high speed ADC IC has separate analog and digital supplies and ground pins for true bipolar operation. Sampling rate can be as high as 25 MHz depending upon clock frequency. The clock input is a CMOS inverter operating from and with logic input levels determined by the Vcc supplies. Care must be taken to keep digital signals away from the analog input and digital ground currents away from the analog ground.—"CA3304, CA3304A CMOS Video Speed 4-Bit Flash Analog to Digital Converter," File #1790, GE/RCA Solid State.

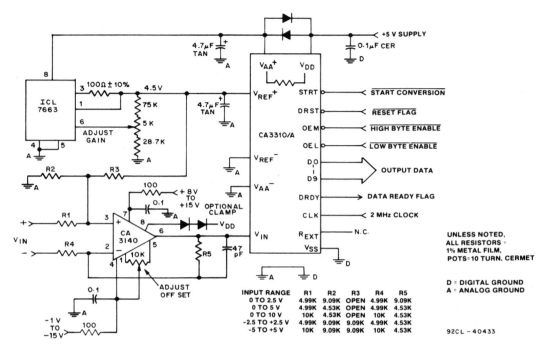

INPUT RANGE	R1	R2	R3	R4	R5
0 TO 2.5 V	4.99K	9.09K	OPEN	4.99K	9.09K
0 TO 5 V	4.99K	4.53K	OPEN	4.99K	4.53K
0 TO 10 V	10K	4.53K	OPEN	10K	4.53K
-2.5 TO +2.5 V	4.99K	9.09K	9.09K	4.99K	4.53K
-5 TO +5 V	10K	9.09K	9.09K	10K	4.53K

UNLESS NOTED,
ALL RESISTORS =
1% METAL FILM,
POTS=10 TURN, CERMET

D = DIGITAL GROUND
A = ANALOG GROUND

92CL - 40433

Analog-to-Digital Converter With Differential Input. Heart of the system is a CA3310 10-bit ADC IC with a voltage reference, input amplifier, and input scaling resistors for several input ranges. An ICL7663 regulator was selected as the reference. The CA3140 op amp provides good slewing capability for high bandwidth input signals. With a 2-MHz clock signal, the maximum input bandwidth will be near 75 kHz. R1 through R5 should be very well matched.—"CA3310, CA3310A CMOS 10-Bit Analog To Digital Converter with Internal Track and Hold," File #1851, GE/RCA Solid State.

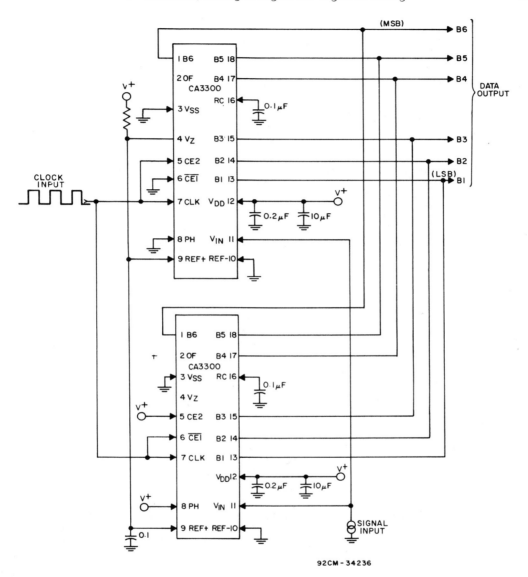

92CM-34236

Analog-to-Digital Converter With 6-Bit Resolution. Circuit uses two CA3300 high speed analog-to-digital converter ICs in parallel to provide a sampling rate of 30 MHz. Supply voltage can range from 3–10 V; clock frequency depends upon sampling rate desired. Accuracy is to within ±0.5 LSB. —"CA3300 CMOS Video Speed 6-Bit Flash Analog to Digital Converter," File #1316, GE/RCA Solid State.

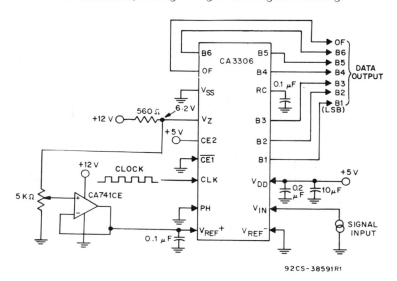

92CS-38591R1

Analog-to-Digital Converter With +5-V Supply Circuit uses CA3306 ADC IC to achieve 6-bit resolution at a sampling rate of 15 MHz. Clock-signal frequency depends upon sampling rate desired. Sampling speed may be doubled by using two CA3306 devices in parallel. Power supply voltage may range from 3–7.5 V—"CA3306 CMOS Video Speed 6-Bit Flash Analog to Digital Converter," File #1789, GE/RCA Solid State.

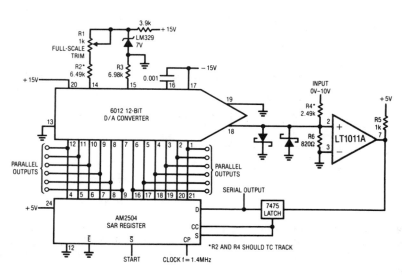

12-Bit Analog-to-Digital Converter. Circuit offers a 10 μs response with both parallel and serial outputs. LT1011A voltage-comparator IC is used because it introduces a very small error. Clock signal to the AM2504 is 1.4 MHz.—"LT1011/LT1011A Voltage Comparator," data sheet, Linear Technology Corporation.

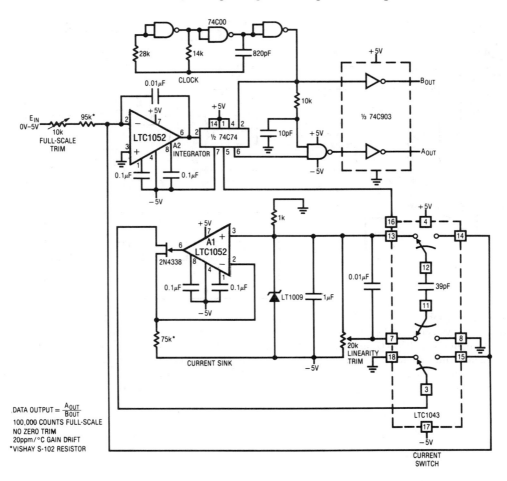

16-Bit Analog-to-Digital Converter. Data output is equal to A-out divided by B-out with 100,000 counts full scale. Shaded devices are LTC1052 chopper-stabilized operational amplifiers. Input signal can range from 0–5 V. The 95-K resistor marked with an asterisk is a Vishay S-102.—"LTC1052/ LTC7652 Chopper Stabilized Operational Amplifier (CSOA)," data sheet, Linear Technology Corporation.

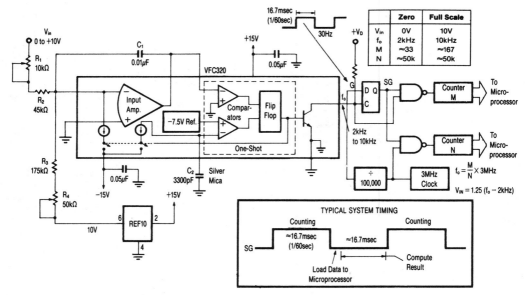

	Zero	Full Scale
V_{in}	0V	10V
f_o	2kHz	10kHz
M	≈33	≈167
N	≈50k	≈50k

$f_o = \frac{M}{N} \times 3MHz$

$V_{IN} = 1.25 (f_o - 2kHz)$

TYPICAL SYSTEM TIMING

Analog-to-Digital Converter Using Voltage-to-Frequency Conversion. VFC320 IC converts a 0–10-V input to a 2–10-KHz output by offsetting the input signal with a reference voltage. The 10-V reference sets a constant input current through R3 and R4 which is added to the signal input current through R1 and R2. A D flip-flop, labeled "SG," is clocked with the VFC320 output. The D input is the output of a 3-MHz clock circuit divided by 100,000. The Q output of the flip-flop serves as the input of two other counters to produce a digital output.—"Voltage to Frequency Converters Offer Useful Options in A/D Conversion," AN-130, Burr-Brown.

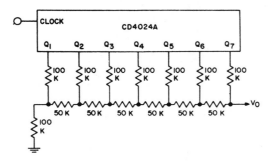

Analog-to-Digital Conversion Using Resistor Ladder Network. CD4024, seven-stage, ripple-carry, binary counter IC is used with a R-2R resistor ladder network which can divide an input ranging from 0–12.8 V into discrete 100-mV increments.— "COS/MOS LSI Counter and Register Design and Applications," ICAN-6166, GE/RCA Solid State.

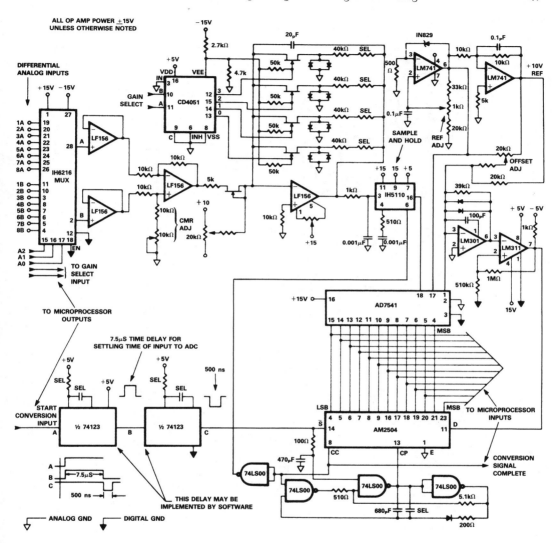

12-Bit Analog-to-Digital Conversion System. Includes complete timing and control features necessary to interface to a microprocessor. The front end is configured differentially using a dual eight-input multiplexer (IH6216) and three LM156 op amps. Following this is a programmable gain stage with a low pass filter at the output. The output goes to the IH5110 sample-and-hold amplifier. The output of the IH5110 is connected to the comparator input (negative input LM301) through the internal 10-K feedback resistor of the 7541 multiplying DAC.—"A Cookbook Approach to High Speed Data Acquisition and Microprocessor Interfacing," A020, GE/Intersil.

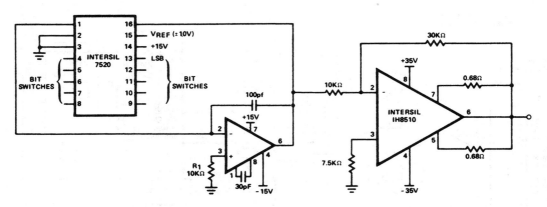

Power Digital-to-Analog Converter. IH5810 power operational amplifier is used with a 7520 monolithic DAC IC to provide drive for DC servomotors, linear and rotary actuators, X-Y printer motors, and similar devices from a digital input. The IH8510 can deliver over 1 A of continuous output and can be operated with ± 35 V power supplies. Accuracy of the output is better than 0.5 LSB and resolution is better than 10 bits.—"Power D/A Converters Using the IH8510," A021, GE/Intersil.

7
Converters, Miscellaneous

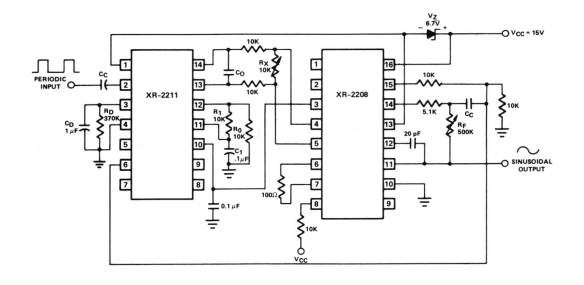

R_X = Distortion Adj. Potentiometer
R_F = Output Amplitude Adj. Pot.

C_C = Coupling Capacitor
($\geqslant 0.1\ \mu$F)

Universal Sine-Wave Converter. Circuit takes any periodic input signal waveform, such as a square wave, and converts it into a low distortion sine wave whose frequency is identical to the repetition rate of the periodic input signal. This circuit is also known as a tracking regenerative filter. Circuit is implemented using a XR-2211 PLL tone decoder and a XR-2208 multiplier IC. The value of capacitor Co, in μF, depends on the output frequency.—"A Universal Sine Wave Converter Using the XR-2208 and the XR-2211," AN-11, Exar Corporation.

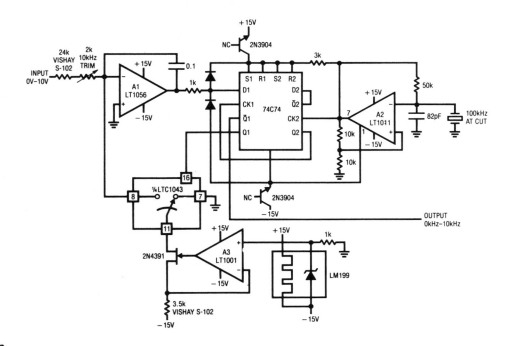

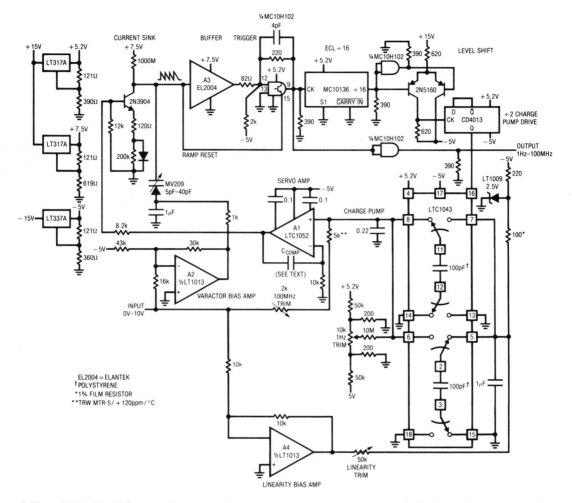

1-Hz to 100-MHz Voltage-to-Frequency Converter. Input voltage from 0–10 V can produce an output signal from 1 Hz to over 100 MHz. Dynamic range is over 160 dB (in eight decades) with 0.06% linearity. ECL counters and flip-flops enable this circuit to operate at an extremely high speed.—"Designs for High Performance Voltage to Frequency Converters," Application Note 14, Linear Technology Corporation.

◄ **Crystal-Controlled Voltage-to-Frequency Converter.** 100-kHz crystal keeps drift of VFC lower than those using other methods of controlling oscillation. Output of 100-kHz oscillator is divided by the flip-flop to produce a 50-kHz clock signal. Circuit can accept an input signal from 0–10 V and produces an output 0–10 kHz.—"Designs for High Performance Voltage to Frequency Converters," Application Note 14, Linear Technology Corporation.

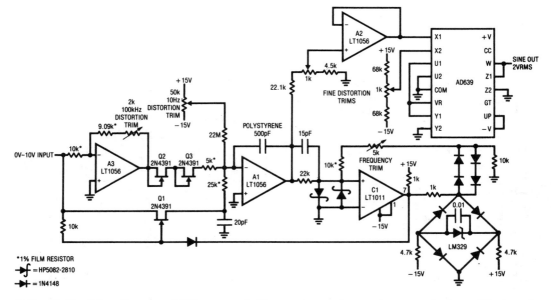

VFC with Sine-Wave Output. VFC accepts a 0–10 V input and delivers an output from 1 Hz to 100 kHz at 2 V RMS. The output is a low distortion sine wave, unlike most other VFC designs whose outputs are generally pulse or square wave. Circuit response to input changes is extremely fast.—"Designs for High Performance Voltage to Frequency Converters," Application Note 14, Linear Technology Corporation.

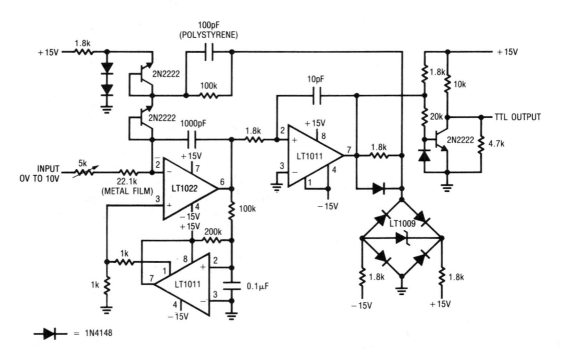

10-Hz to 1-MHz Voltage-to-Frequency Converter. Shaded device can be either a LT1056 or a LT1022 op amp. An input voltage from 0–10 V will produce an output frequency from 10 Hz to 1 MHz. Output is TTL-compatible.—"LT1022 High Speed Precision JFET Input Operational Amplifier," data sheet, Linear Technology Corporation.

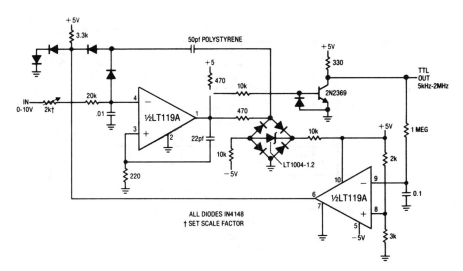

5-kHz to 2-MHz Voltage-to-Frequency Converter. Uses both halves of LT119A dual comparator (LM119 may also be used). Accepts an input between 0–10 V and produces a TTL-compatible output from 5 kHz–2 MHz. The 2-K potentiometer at input is used to calibrate the circuit.—"LT119A/LT319A Dual Comparator," data sheet, Linear Technology Corporation.

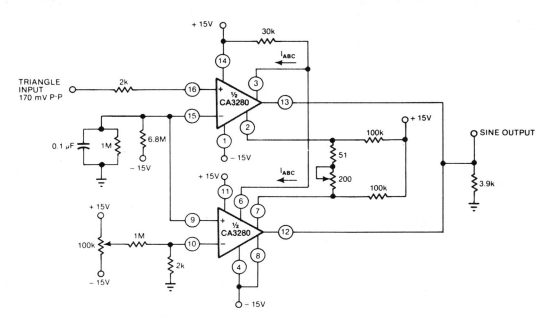

Triangle to Sine-Wave Converter. Two 100-K resistors connected between the differential-amplifier emitters of each half of the CA3280 op amp and the positive supply voltage reduce current flow through the differential amplifier. This permits the amplifier to cut off fully during peak-input signal excursions. Harmonic distortion of this circuit is 0.37% or less.— "Dual Variable Op Amp IC, the CA3280, Simplifies Complex Analog Designs," ICAN-6818, Ge/RCA Solid State.

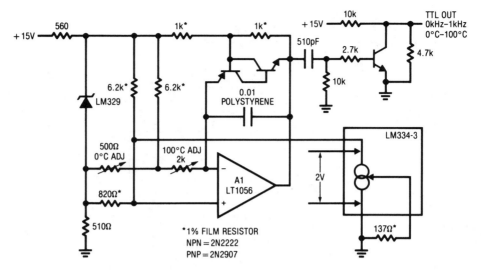

Temperature-to-Frequency Converter. Produces a TTL-compatible output ranging from 0–1 kHz for a temperature reading from 0–100 degrees Celsius (C). Heart of circuit is a LM334 temperature sensor whose output serves as part of the non-inverting input to a LT1056 op amp. Calibration is accomplished with temperatures of 0 and 100 degrees Celsius; the poten-tiometer labeled 0 degrees Celsius is adjusted for an output of 0 Hz at 0 degrees C while the potentiometer labeled 100 degrees C is adjust for a 1 kHz output at that temperature.—"Some Techniques for Direct Digitization of Transducer Outputs," Application Note 7, Linear Technology Corporation.

8

Digital and Logic Circuits

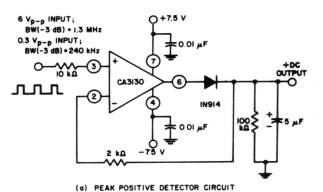

(a) PEAK POSITIVE DETECTOR CIRCUIT

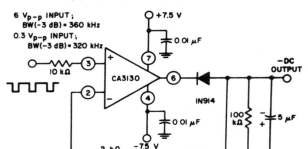

(b) PEAK NEGATIVE DETECTOR CIRCUIT

Peak Detectors.   CA3130 BiMOS operational devices are used in configurations for detection of positive and negative signal peaks. The bandwidth of the peak negative detector is less than that of the positive peak detector.— "CA3130 BiMOS Operational Amplifiers," File #817, GE/ RCA Solid State.

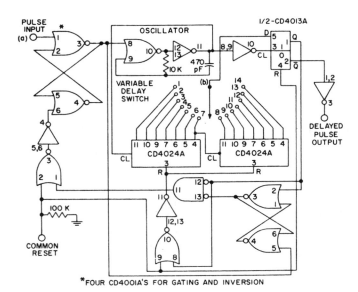

*FOUR CD4001A'S FOR GATING AND INVERSION

Pulse Delayer.   CD4024 seven stage ripple-carry binary counter is used to derive a desired pulse-output delay time. Various delay times can be obtained by detecting different CD4024 outputs. Extremely long and accurate delay times can be realized by using an accurate oscillator and multiple CD4024 devices.—"COS/MOS MSI Counter and Register Design and Applications," ICAN-6166, GE/RCA Solid State.

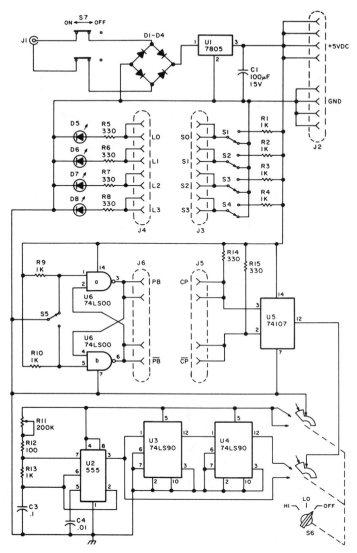

IC Digital Trainer.   Designed to be a classroom or self-study aid for learning the principles of digital-logic devices. The four SPDT toggle switches, S1–S4, can select between + 5 V DC and ground to provide a logic 0 or 1 at jacks S0–S3. S5 is debounced by a NAND gate and gives an output and its complement at PB and $\overline{PB}$. LEDs D5–D8 give visual indication of logic levels. D1–D4 are 1N4001 or equivalent. Original article gives complete details on the layout and construction of the trainer.—R. Need, "Build a Digital Logic Trainer," 73 *Magazine*, March 1986, pp. 40–43.

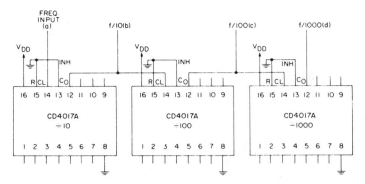

Multi-Decade Frequency Division. Three CD4017 decade counter/divider ICs are used to provide division of input signal by 10, 100, and 1000. Decimal display may be added if desired.—"COS/MOS MSI Counter and Register Design and Applications," ICAN-6166, GE/RCA Solid State.

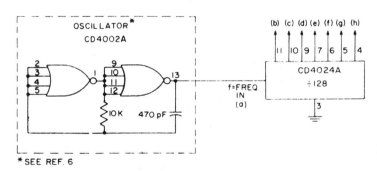

* SEE REF. 6

Binary Frequency-Divider. CD4024 seven-stage ripple-carry binary counter IC is used to divide output of oscillator by a factor of 2–128. Multiple CD4024 devices can be "stacked" for extra frequency-division.—"COS/MOS MSI Counter and Register Design and Applications," ICAN-6166, GE/RCA Solid State.

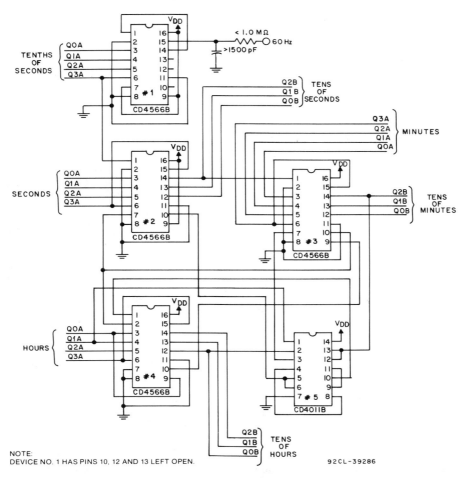

Twelve-Hour Clock. Circuit uses four 4566, industrial time-base generator ICs and a 4011 quad NAND gate. Each 4566 consists of a divide-by-10 ripple-counter and a divide-by-six ripple-counter which permit stable time-generation from a 60-Hz clock signal.—"CD4566B Types CMOS Industrial Time-Base Generator," File #1728, GE/RCA Solid State.

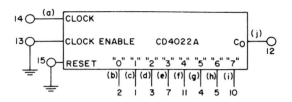

Divide-by-Eight Counter/Decoder. CD4022 divide-by-eight counter IC provides both counting and decoding functions.—"COS/MOS MSI Counter and Register Design and Applications," ICAN-6166, GE/RCA Solid State.

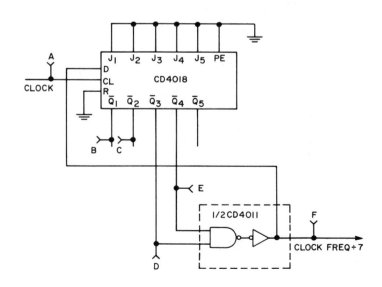

Divide-by-Seven Counter. 4018 divide-by-N counter IC is configured for divide-by-seven operation.— "Design of Fixed and Programmable Counters Using the RCA CD4018 COS/MOS Presettable Divide by N Counter," ICAN-6498, GE/RCA Solid State.

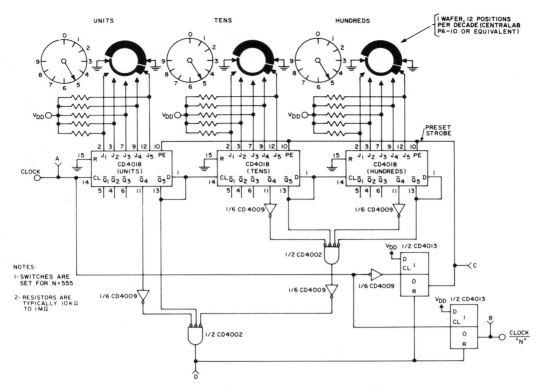

2–999 Three-Decade Counter. Selector switches determine the division ratio of the three 4018 counter ICs. Extension to higher ranges is readily accomplished by using additional 4018 devices. Switches are arranged so that switch position nine is equivalent to zero, position eight to one, and so forth.—"Design of Fixed and Programmable Counters Using the RCA CD4018 COS/MOS Presettable Divide by N Counter," ICAN-6498, GE/RCA Solid State.

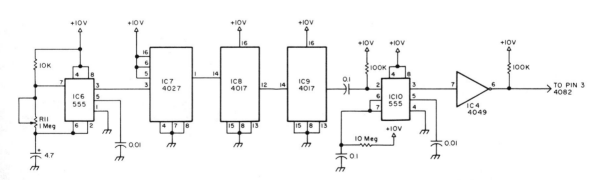

Clock-Signal Generator. Potentiometer R11 (1M) can be adjusted to provide a high-to-low logic clock-pulse every 2–90 minutes. The first 555 timer, IC6, is configured as an astable oscillator whose output frequency is controlled by R11. The 4027, IC7, produces a perfectly symmetrical square wave signal for IC8, the 4017 divide-by-10 counter. The output of IC8 is further divided by IC9, a second 4017. The output of the second 4017 is fed to IC10, a second 555, which is used as a monostable multivibrator. The 4049 hex buffer IC (IC4) inverts the output of the second 555. Circuit was originally designed to control the battery recharging cycle in a solar power system.—M. Bryce, "Total Solar," 73 *Magazine*, May 1986, pp. 60–64.

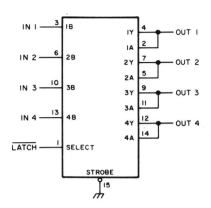

4-Bit Transparent Latch.   Formed from 74157 data selector, latch is formed by connecting the outputs to their respective "A" inputs and applying data to the "B" inputs. When the select input is high, the "B" input data appears on the outputs and on the "A" inputs. When the select input is low, the output data is held via the "A" inputs until the select line again goes high.—P. Selwa, "4-Bit Transparent Latch," 73 *Magazine*, August 1985, p. 67.

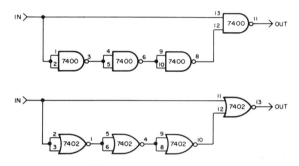

Pulse-Edge Detectors.   Circuit using 7400, TTL, quad NAND gate IC will produce a single, low pulse when the input goes from low to high. Circuit using the 7402 TTL quad NOR gate will produce a single, high pulse when the input switches from high to low.—P. Selwa, "Two-Way Edge Detection," 73 *Magazine*, August 1985, p. 67.

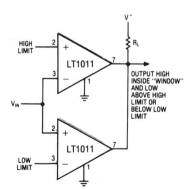

Window Detector.   Dual, LT1011 voltage comparator ICs are used to indicate range-of-input voltage. Output of circuit is high when input voltage is within the "window" and low when outside it.—"LT1011/LT1011A Voltage Comparator," data sheet, Linear Technology Corporation.

Magnetic-Transducer Detector. Produces a TTL-compatible output from a magnetic pickup or transducer when a +5-V supply is used. The IC is a LT311A voltage comparator; a LM311 may be substituted.—"LT111A/LT311A Voltage Comparator," data sheet, Linear Technology Corporation.

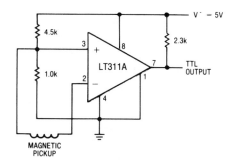

Window Detector with TTL-Compatible Output. Both halves of a LT319A dual comparator are used to produce a window comparator whose output can directly drive TTL inputs. The output voltage is high (5 V) if Vin is less than or equal to Vut yet is greater than or equal to Vlt; otherwise, output is low (0 V). The allowed window for a single +5-V supply is 1.2–3.8 V. A LM319 may be substituted.—"LT119A/LT319A Dual Comparator," data sheet, Linear Technology Corporation.

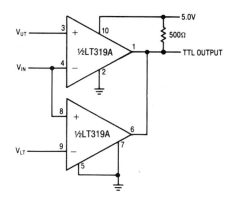

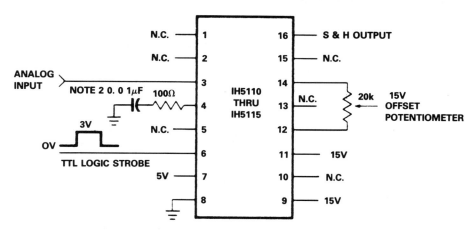

TTL-Controlled Sample and Hold. The IH5110/5115 monolithic sample and hold IC samples an analog input when the TTL logic input (pin 6) is high and holds it when the TTL input is low. The 20-K offset potentiometer is adjusted to set the output to zero in the absence of an input signal.—"IH5110-IH5115 General Purpose Sample and Hold," data sheet, GE/Intersil.

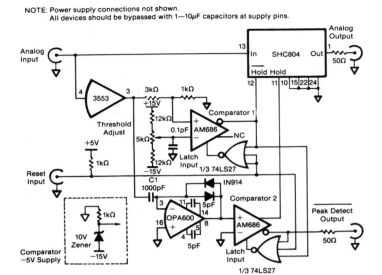

NOTE: Power supply connections not shown.
All devices should be bypassed with 1—10µF capacitors at supply pins.

High-Speed Peak-Detector. Captures and holds the values of fast-changing analog signals; acquisition time is 350 ns maximum with a settling time of 150 ns maximum. The occurrence of a peak is detected by the diode-clamped differentiator circuit built around a OPA600 op amp. When the input voltage peaks and begins to decrease, the current in C1 (1000 pF) reverses. This causes the OPA600 to switch rapidly from its negative to positive output state. Comparator 2 detects this change and switches the SHC804 into a "hold" mode, freezing the peak value.—"A High Speed Peak Detector," AN-137, Burr-Brown.

9
Display Circuits

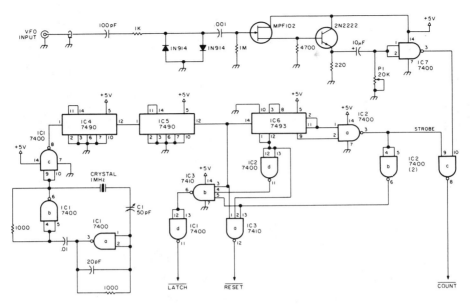

Universal Digital-Frequency Display. Designed for use with most transceivers and receivers operating from 1.5–30 MHz. Omits the MHz reading and counts instead the output of the transceiver or receiver VFO, displaying it in kHz. The circuit covers 500-kHz segments; S1 selects whether the count begins at 000 kHz or 500 kHz. S2 offsets the frequency count by 3 kHz in an up-to-down direction when SSB signals are being tuned. Circuit is based on the 74192 counter ICs, which allow either up-or-down counting. Signal input is taken from the transceiver or receiver VFO circuit and the unit is aligned to tune the VFO output; original article gives details on this procedure. The time-base oscillator signal controlled by the 1-MHz crystal must be precisely adjusted (preferably with an accurate frequency-counter) for accurate frequency-readout. LS versions of TTL ICs should be used to minimize heat generation. Original article includes construction details. While not as accurate as other methods of digital-frequency display, this circuit is significantly less expensive.—T. Miller, "What You See is Where You're At," 73 *Magazine*, May 1986, pp. 54–59.

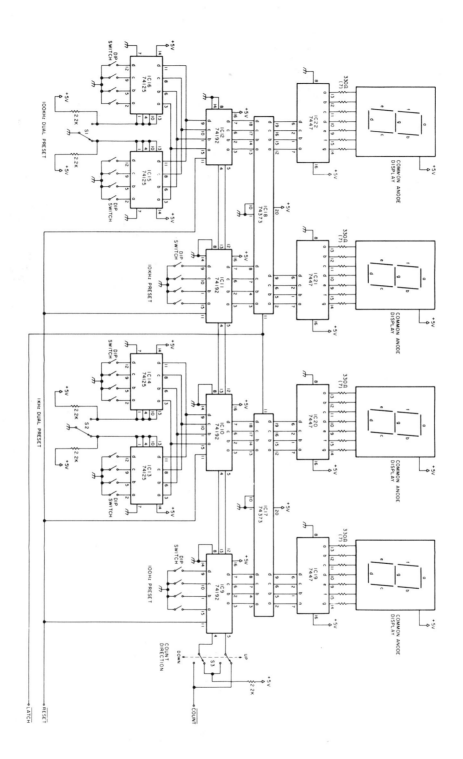

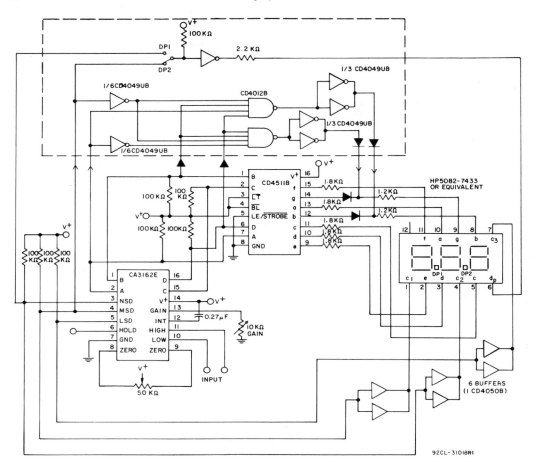

92CL-31018R1

Common Cathode-LED Display. CA3162E ADC IC is connected to a 4511 decoder/driver IC to operate a common cathode-LED display. The 4511 remains blank for all BCD codes greater than nine. After 999 mV, the display becomes blank rather than displaying "EEE." When displaying negative voltage, the first digit remains blank instead of displaying "−," and the display becomes blank during positive or negative overrange.—CA3162E, CA3162AE A/D Converter for 3-Digit Display," File #1080, GE/RCA Solid State.

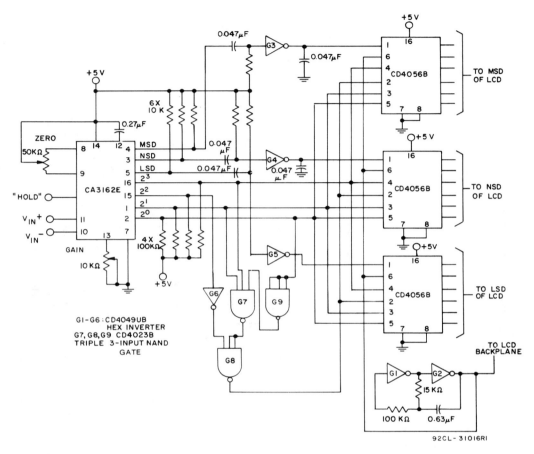

92CL-31016RI

LCD System. CA3162E monolithic ADC IC is heart of system to drive LCDs. Multiplexing of LCD digits is not practical, since LCDs must be driven by an AC signal and the average voltage across each segment is zero. Therefore, three CD40568 LCD decoder/driver ICs are used; each 4056 contains an input latch so that the BCD data for each digit may be latched onto the decoder using as strobes the inverter digit-select outputs of the CA3162E. Inverters G1 and G2 are used as an astable multivibrator to provide the AC drive to the LCD backplane.— "CA3162E, CA3162AE A/D Converter for 3-Digit Display," File #1080, GE/RCA Solid State.

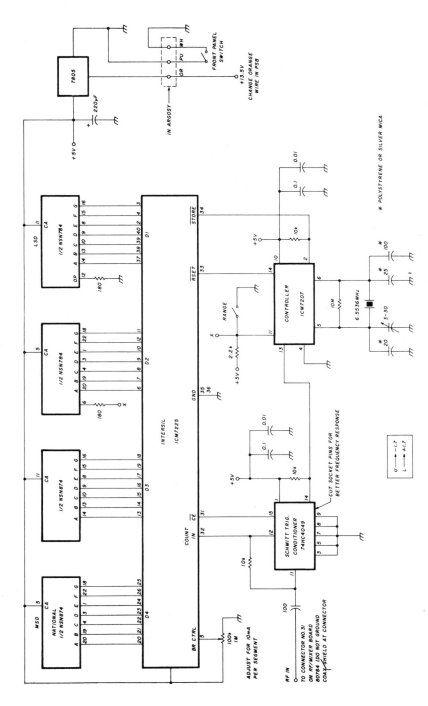

Receiver Digital-Frequency Counter and Display. Designed for use with Ten-Tec "Argosy" SSB transceiver which uses a 9-MHz IF section and covers 2.5–29.7 MHz. The RF injection-signal is fed to conditioning IC 74HC4049 via a coaxial cable and a 100-pF capacitor. National NSN874 common-anode displays were selected because of space restrictions of the Argosy transceiver. Pins on the 74HC4049 and ICM7207 may have to be shortened for best results. Original article gives full details on construction, alignment, and installation.—C. Drentea, "Upgrading the Ten-Tec Argosy," *ham radio*, November 1986, pp. 38–50.

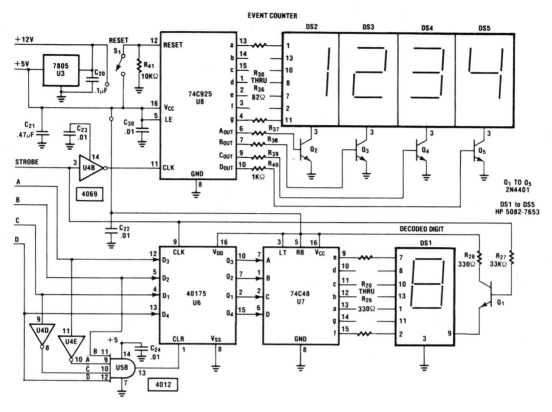

DTMF-Decoder Display-and-Counting Circuit.
Detects valid digit decoding by a DTMF decoder and displays digit decoded. Circuit consists of a valid-digit counter made up of four seven-segment LED displays, which are driven by a 74C925 counter and a single, seven-segment LED display and four ICs. Special characters will appear distorted when displayed.—"Using the S3525A/B DTMF Bandsplit Filter," applications note, Gould/AMI Semiconductors.

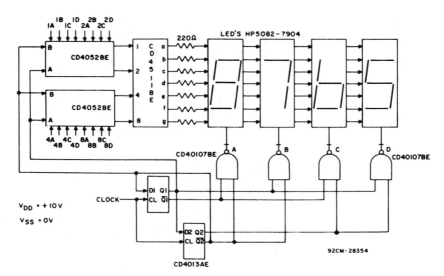

Multiplexed LED-Display Circuit. Data is presented in the form of four BCD numbers to be displayed on the four seven-segment LEDs; the clock determines the multiplexing rate. The two D-type flip-flops of the 4013 are arranged as a two-stage Johnson counter, the two Q-outputs of which select the data transferred by the 4052 multiplexers to the 4511 decoder/driver. The same Q-outputs are decoded by the four NAND buffers and are used to turn on the seven-segment displays in correct sequence.— "Applications of CD40107BE COS/MOS Dual NAND Buffer," ICAN-6564, GE/RCA Solid State.

10

Electro-Optic and Fiber Optic Circuits

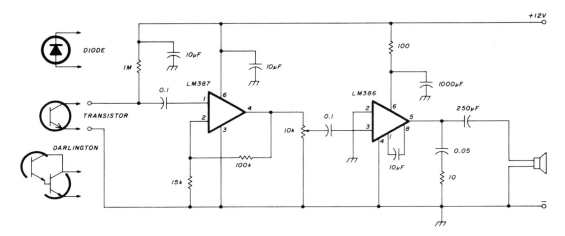

Electro-Optic Receiver. Designed for use with laser-transmission systems. Photo detectors used with this circuit may be photodiodes, bipolar phototransistors, photo Darlington transistors, or PIN photodiodes; best choice is usually the MRD500 or MRD510 photodiodes. Original article includes details on selecting appropriate lasers, modulating techniques, and operating precautions.—S. Noll, "Communicating on 474,083 GHz," *ham radio*, December 1986, pp. 10–22.

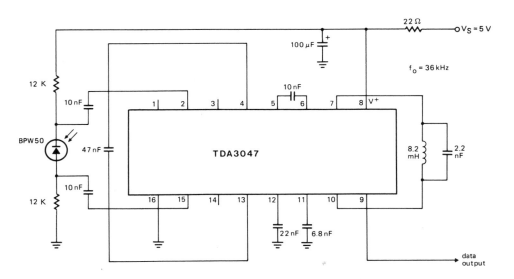

Wide-Band Infrared Receiver. TDA3047 device is a single-chip infrared receiver and demodulator designed for input signals which are typically AM at 36 kHz. For better sensitivity, both 12-K resistors may have higher values, but the circuit may be prone to instability. Outputs of the circuit are symmetrical square waves.—"TDA3047 I/R Preamplifier," 1985 Linear LSI Data and Applications Manual, 5-126, Signetics Corporation.

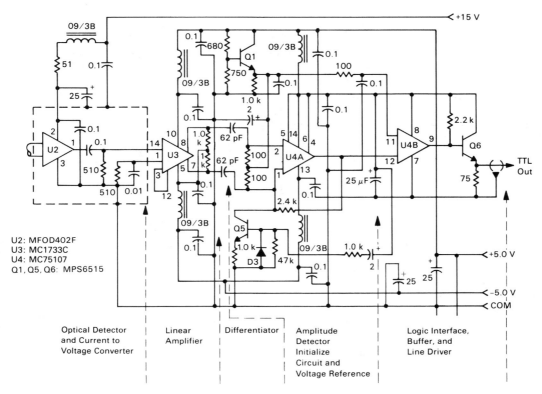

U2: MFOD402F
U3: MC1733C
U4: MC75107
Q1, Q5, Q6: MPS6515

Optical Detector	Linear	Differentiator	Amplitude	Logic Interface,
and Current to	Amplifier		Detector	Buffer, and
Voltage Converter			Initialize	Line Driver
			Circuit and	
			Voltage Reference	

Fiber Optic Receiver. Capable of 20-megabaud data rate and has TTL-compatible outputs. When used with matching transmitter, it may be integrated into a complete fiber optic transceiver or used for full duplex operation. The MFOD402F performs both the optical detection and current to voltage conversion functions. The MC1733 serves as a linear am-plifier for the detected and converted signal. Original applications note includes construction and oper-ating data along with PC board artwork.—"A 20 Mbaud Full Duplex Fiber Optic Data Link Using Fiber Optic Active Components," AN-794, Moto-rola, Inc.

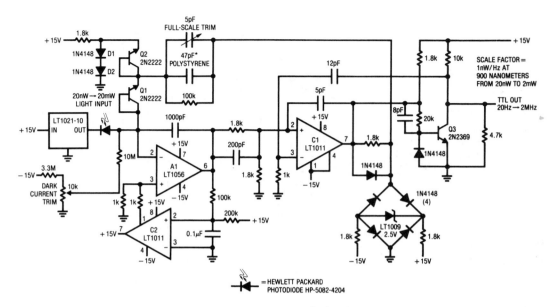

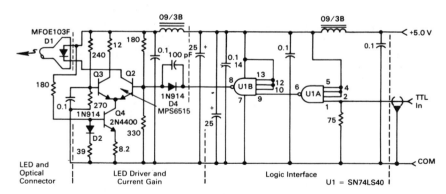

Photodiode Digitizer. Directly converts a photodiode's output current into an output frequency with 100 dB of dynamic range. Optical input power of 20 nW to 20 mW produces a linear, calibrated 20 Hz to 1 MHz output. Output response to input light-steps is fast and cost is low. The photodiode's output current feeds a highly modified, high frequency version of a Pease charge-pump current-to-frequency converter.—"Some Techniques for Direct Digitization of Transducer Outputs," Application Note 7, Linear Technology Corporation.

Fiber Optic Transmitter. Capable of 20-megabaud data rate and has TTL-compatible inputs. When used with matching receiver, system may be integrated as a complete transceiver and used for full duplex operation. Two sections of a 74LS40 dual, four-input, NAND gate are used in cascade in the logic-interface section. Input impedance is 75 ohms and a coaxial cable is suggested. Original applications note includes full construction details, circuit theory, and PC board artwork.—"A 20 Mbaud Full Duplex Fiber Optic Data Link Using Fiber Optic Active Components," AN-794, Motorola, Inc.

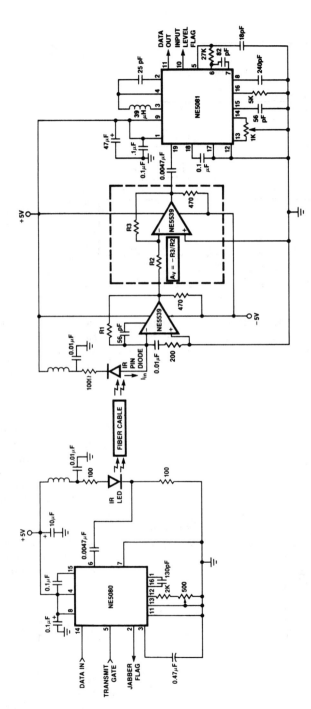

Simplex Fiber Optic System. NE5080 FSK transmitter IC and NE5081 FSK receiver IC are used as a simplex fiber optic link with a center frequency of 5 MHz. Since the NE5081 can adequately accept signals below 10 mV, the gain stage (enclosed in dashed lines) may be eliminated if the attenuation in the link is low. Original article gives details on setup and alignment.—"Applications Using the NE5080, NE5081," AN195, Signetics Corporation.

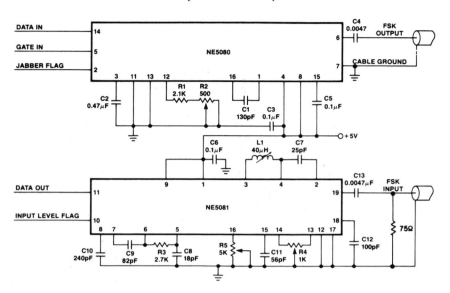

2-Megabaud, Full-Duplex Modem. NE5080 FSK transmitter IC and NE5081 FSK receiver IC are used together to produce a FSK modem offering up to a two-megabaud speed at a 5-MHz carrier frequency. Dynamic range is 30 dB, allowing attachment to various points in a network without any gain adjust-ment. Original article gives details on setup proce-dures and permissible baud rates for a given carrier frequency and type of coaxial cable.—"Applications Using the NE5080, NE5081," AN195, Signetics Corporation.

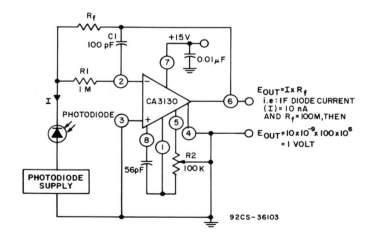

Photodiode Current-to-Voltage Converter. CA3130 is used to amplify input currents which can be in the sub-picoampere region. Circuit shown provides a ground-referenced output voltage that is proportional to the current flowing through a photodiode in the input circuit. R1 is used to limit input current to a safe value in case the back-biased photodiode should "avalanche" and expose the input of the CA3130 to the comparatively high voltages sometimes used in photodiode power supplies.—"Understanding and Using the CA3130, CA3130A, and CA3130B BiMOS Operational Amplifiers," ICAN-6386, GE/RCA Solid State.

11
Microprocessor and Microcomputer Circuits

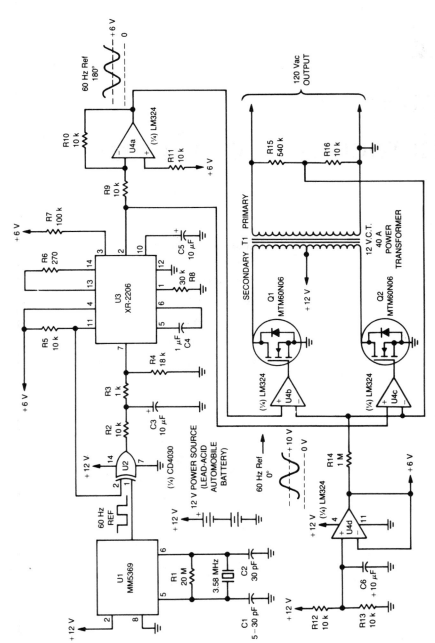

Uninterruptible Power Supply for Personal Computers. Delivers 120-V AC at 60 Hz and 4 A with a backup power source provided by a 12-V lead-acid auto battery. This can provide several minutes of backup power if the main 120-V AC source fails. A crystal-controlled 60-Hz time base allows any computer real-time clock to maintain its accuracy. Two MTM60N06 power FETs alternately switch current through a center-tapped filament transformer with its primary and secondary reversed. The 120-V output is compared with a 60-Hz reference in a closed-loop configuration that maintains a constant output at optimum efficiency.—B. Williams, "Uninterruptible Power Supply for Personal Computers," *TMOS Power FET Design Ideas*, p. 43, Motorola, Inc.

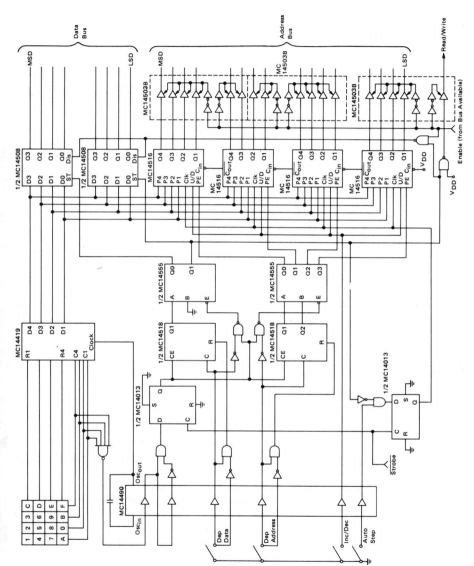

CMOS Keyboard Data-Entry System. Can be used with any CMOS-compatible memory system with external control of an address bus of up to 16 bits and a data bus of up to 8 bits. The 4 × 4 keypad switch-array is coded in hexadecimal and allows each keystroke to code 4 binary bits. Primary application is with 6800-family microprocessors.—"A CMOS Keyboard Data Entry System for Bus Oriented Memory Systems," AN-759, Motorola, Inc.

83

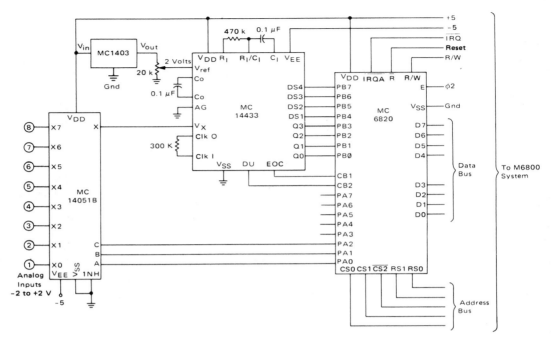

Eight-Channel Data-Acquisition System. Designed to be used with the 6800-series microprocessor family, circuit consists of a MC14433 CMOS analog-to-digital converter, MC1403 2.5-V reference, MC14051B analog multiplexer, and a MC6820 programmable interface-adapter. Original applications note includes a program listing for the MC6820. The conversion technique used by the MC14433 is a modified dual ramp featuring auto-zero, auto-polarity, and high input impedance.— "Data Acquisition Networks with NMOS and CMOS," AN-770, Motorola, Inc.

Bicycle Computer. MC146805G2 single-chip microcomputer serves as "brain" of bicycle computer which can compute instantaneous speed, average speed, and cadence (pedal crank revolutions per minute). Circuit also offers a resettable trip and long distance odometer and a selection of English or metric units. The computer may also be calibrated for wheel size. The LCD is a FE0201 or equivalent. Original applications notes includes a software listing for the MC146805G2, construction and installation data, and PC-board artwork.—"Bicycle Computer Using the MC146805G2( )1 Microcomputer," AN-858, Motorola, Inc.

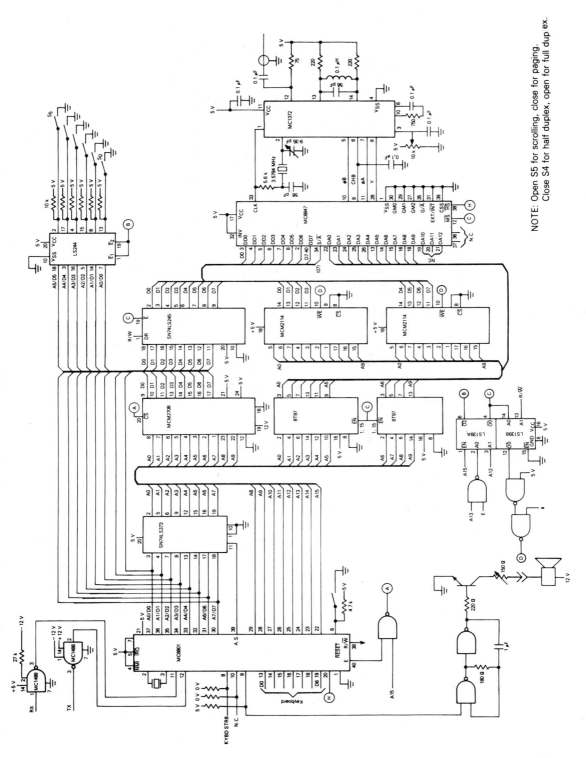

NOTE: Open S5 for scrolling, close for paging.
Close S4 for half duplex, open for full dup ex.

87

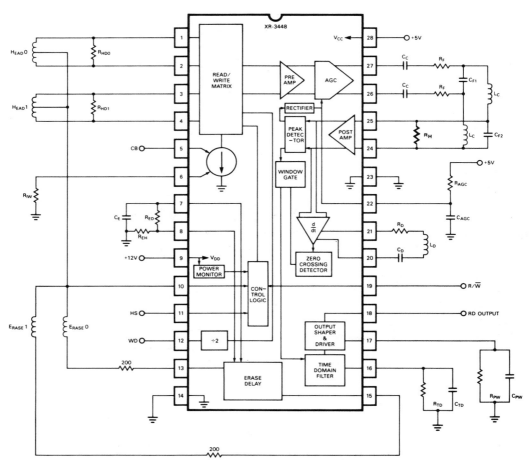

Floppy Disk Read/Write Controller. XR-3448 floppy disk read/write IC is used in circuit providing all read and write functions for double-sided floppy disks in 8-, 5.25-, and 3.5-in. formats. Circuit will work with single- or dual-head systems. Accompanying table gives typical parts values.—"XR-3448 Floppy Disk Read/Write," data sheet, Exar Corporation.

TYPICAL COMPONENT VALUES

Component	Typical Value	Recommended Range	Component	Typical Value
R_{IW}	560Ω	180 - 680Ω	R_H	1.5kΩ
R_{EO}, R_{E1}	280Ω	100 - 680Ω	C_C	.022μF
R_{ED}, R_{EH}	10kΩ	5k - 30kΩ	R_F	470Ω
C_E	.047μF	0.01 - 0.068μF	C_{F1}	.001μF
R_{PW}	10k	5k - 25kΩ	C_{F2}	470pF
C_{PW}	100 pF	51 pF - 1000 pF	L_C	680μH
C_{TD}	100pF	51 pF - 330 pF	R_D	200Ω
R_{AGC}	6.81k	3.3k - 25kΩ	C_D	1000pF
C_{AGC}	.1μF	.01μF - 1μF	L_D	56μH

12
Miscellaneous Circuits

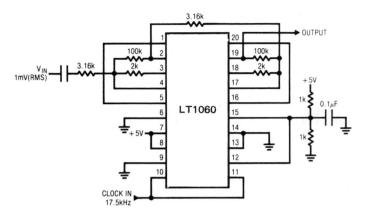

175-Hz Bandpass Filter with Gain of 1000. Fourth-order bandpass filter built around LTC1060 dual-switched capacitor filter IC offers a gain of 1000 from a single +5-V supply. Filter has a peak response at 175 Hz with a gain of over 60 dB at that frequency and drops below unity gain at frequencies below 125 Hz and over 225 Hz. Input clock-signal may be from a TTL or CMOS source.—"LTC1060 Universal Dual Filter Building Block," data sheet, Linear Technology Corporation.

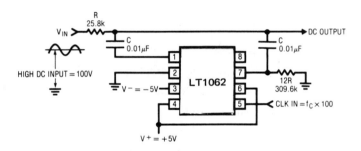

AC-Signal Filter for High DC Voltages. Acts as a fifth-order low-pass filter for unwanted AC signals "riding" on the DC input. The cutoff frequency of the filter is a function of the clock input signal; the clock frequency is 100 times that of the desired cutoff frequency. For example, a clock frequency of 100 kHz produces a cutoff frequency of 1 kHz.— "LTC1062 5th Order Low Pass Filter," data sheet, Linear Technology Corporation.

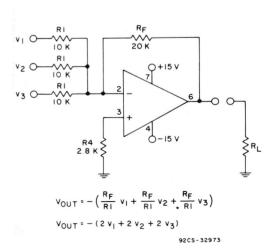

$$V_{OUT} = -\left(\frac{R_F}{RI} v_1 + \frac{R_F}{RI} v_2 + \frac{R_F}{RI} v_3\right)$$

$$V_{OUT} = -(2 v_1 + 2 v_2 + 2 v_3)$$

92CS-32973

Summing Amplifier. Uses CA3193 precision BiMOS operational amplifier to produce output voltage which is a product of the input voltages, V1, V2, and V3, as indicated by the equations shown. —"CA3193/, CA3193A/ High Reliability Slash (/) Series BiMOS Precision Operational Amplifiers," File #1730, GE/RCA Solid State.

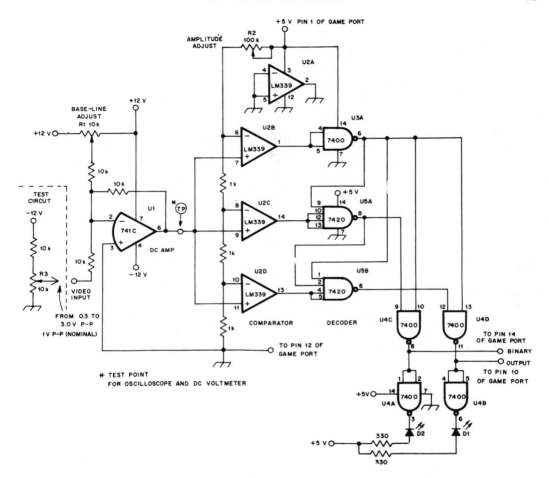

WEFAX Converter for IBM Microcomputers. Converts signals received from weather satellites using the WEFAX standard into a form which can be displayed on IBM PC microcomputers. The WEFAX system transmits photos using amplitude-modulated tones, and the circuit shown is essentially an analog-to-digital converter. The circuit converts the input video signal into four amplitude ranges, assigns a 2-bit binary number to each range, and delivers the number to the microcomputer's computer-game port. As the video input amplitude varies, the microcomputer "sees" one of four binary numbers. Four color pictures can be reproduced if the microcomputer has a color monitor; pictures may be stored on a disk and printed out on dot-matrix printers. Original article includes details on supporting software.—E. Schwittek and W. Schwittek, "WEFAX Pictures on Your IBM PC," *QST*, June 1985, pp. 14–18.

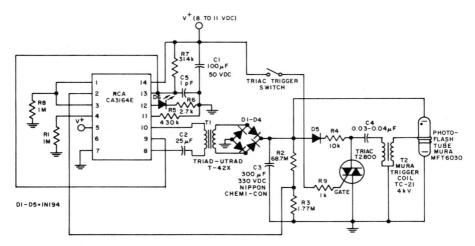

Photoflash Circuit for Cameras. CA3164E BiMOS control IC drives the primary of step-up transformer T1 with symmetrically chopped current at a rate of 500–2000 Hz. Diodes D1–D4 rectify the output of T1 and charge C3 to approximately 280 V. The maximum charge of C3 is determined by the ratio of R2 and R3. A tap between R2 and R3 supplies the turnoff signal for the CA3164E. When C3 is charged, C4 is also charged. When the triac trigger-switch is closed, C4 discharges through the primary of T2, producing a 4-kV pulse in the secondary to drive the photoflash tube.—"CA3164E BiMOS Control Chip Extends Battery Life in Camera's Photoflash Circuit," ICAN-6823, GE/RCA Solid State.

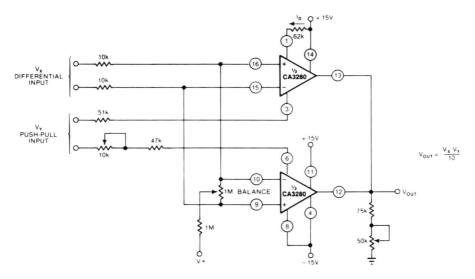

$$V_{OUT} = \frac{V_x V_y}{10}$$

Four-Quadrant Multiplier. Both halves of a CA3280 dual op amp IC are used to produce a multiplier with differential operation. Output voltage is equal to the product of Vx and Vy divided by ten. Higher Vx input voltages can be applied by increasing the input resistor values; Vy must be symmetrical above ground and limited to the supply voltages.— "Dual Variable Op Amp IC, the CA3280, Simplifies Complex Analog Designs," ICAN-6818, GE/RCA Solid State.

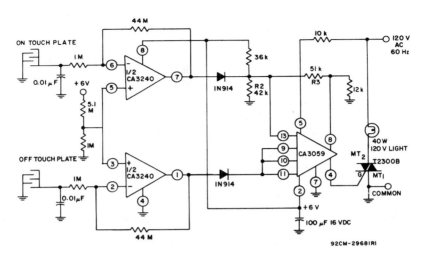

92CM-29681RI

On/Off Touch Switch. CA3240 op amps sense small currents flowing between contact points on a touch plate (a PC board metallization grid). The currents flowing between the "on" and "off" plates control the action of the CA3059 zero-voltage switch IC.

High circuit-impedance prevents shocks.—"Features and Applications of RCA Integrated Circuit Zero Voltage Switches (CA3058, CA3059, and CA3079)," ICAN-6182, GE/RCA Solid State.

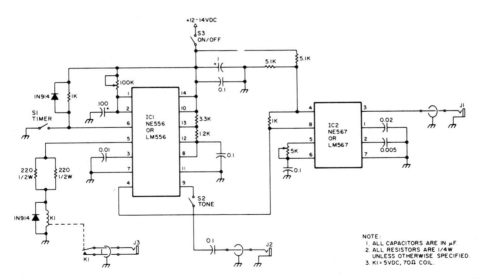

Closed-Loop Cassette-Tape Controller. Used to control a tape recorder with a continuous closed-loop cassette whose output is fed to a transmitter or transceiver. Jack J1 is connected to the tape recorder's earphone jack, J2 to the recorder's microphone input, and J3 to the recorder's auxiliary input jack; the line between J1 and the recorder earphone jack is also wired to the transmitter or transceiver headphone

and microphone jack. Closing switch S1 will cause the message on the cassette tape to be transmitted. An audio tone of 2500 Hz was chosen for the 567 decoder IC since that frequency lies outside of the 200–2400-Hz passband used for audio input to most SSB transmitters. K1 is a reed relay.—D. Malley, "A VOX in a Box," 73 *Magazine*, December 1986, pp. 34–36.

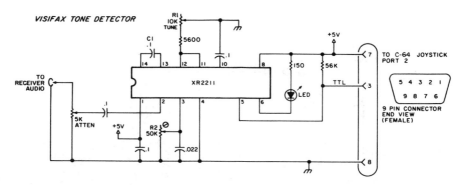

FAX-Tone Detector. Designed for use with a Commodore C-64 microcomputer, circuit detects tones used to make up facsimile (FAX) signals and generates a TTL-level signal corresponding to those tones. Commodore C-64 uses programs which "read" this circuit's output and generate a weather chart or other facsimile material, outputting it to a printer. Supporting software is required for the C-64 or other microcomputer used.—G. Sargent, "Just the FAX, Ma'am," 73 *Magazine*, October 1986, pp. 24–26.

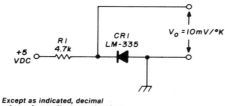

Except as indicated, decimal values of capacitance are in microfarads (μF); others are in picofarads (pF); resistances are in ohms. k = 1,000   M = 1,000,000

Temperature Sensing. Based upon LM335 diode from National Semiconductor, which can measure temperature over a range from -10 to $+100$ degrees Fahrenheit (-23 to $+38$ degrees Celsius). Output across the diode will be 10 mV per degree Kelvin. Kelvin degrees are the same as Celsius degrees except that zero is referenced to absolute zero instead of the freezing point of water; thus 0 degrees C = 273 degrees K.—J. Carr, "Practically Speaking," *ham radio*, April 1986, pp. 75–81.

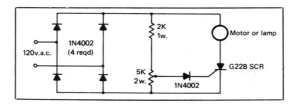

SCR Dimmer. SCR is used to control the duty cycle of the rectified AC resulting in anything from zero to full power being applied to the load. Peak power is always quite high from this circuit. Although designed as a light dimmer, circuit can also be used with some AC motors.—I. Math, "Math's Notes," *CQ*, March 1986, p. 44.

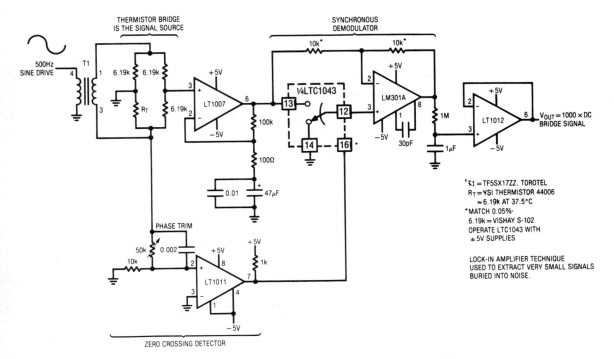

Lock-In Amplifier.   Used to extract very small signals buried in noise through an extremely narrow bandwidth. A thermistor bridge provides the signal source; RT is a YSI thermistor 44006 or equivalent. The 6.19-K resistors are Vishay S-102 or equivalent. Output is applied to a synchronous detector using a LTC1043 instrumentation-grade switched-capacitor building block. Asterisked components must be matched to within 0.05%. Transformer T1 is a Torotel TF5SX17ZZ or equivalent.—"LTC1043 Dual Switched Instrumentation Switched-Capacitor Building Block," data sheet. Linear Technology Corporation.

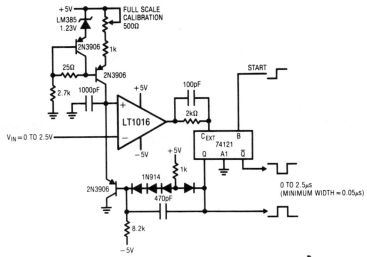

Voltage-Controlled Pulse-Width Generator.   An input voltage between 0–2.5 V produces output pulses ranging in duration from 0.05–2.5 μS. Complementary outputs are available at Q and $\overline{Q}$ as shown. Shaded device is LT1016 comparator.—"LT1016 Ultra Fast Precision Comparator," data sheet, Linear Technology Corporation.

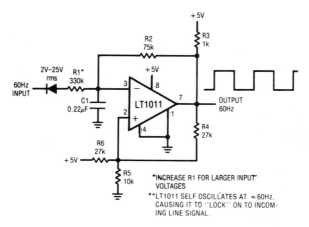

60-Hz Line Synchronizer. LT1011 voltage comparator self-oscillates at approximately 60 Hz, causing it to lock onto incoming-line (60 Hz) signal. Output is 60-Hz square wave.—"LT1011/LT1011A Voltage Comparator" data sheet, Linear Technology Corporation.

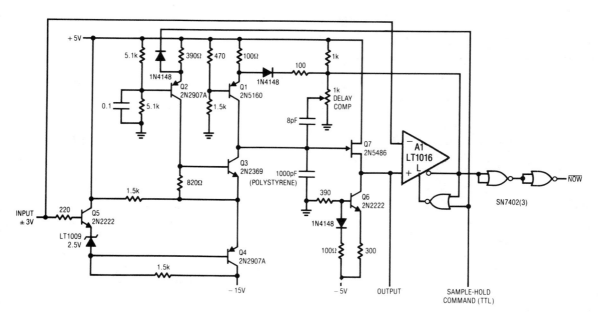

200-ns Sample and Hold. Acquisition time of 200 ns or less is well beyond the capabilities of monolithic sample and hold. When the sample and hold line goes low, a linear ramp starts just below the input level and ramps upward. When the ramp voltage reaches the input voltage, the LT1016 comparator shuts off the ramp, latches itself off, and sends out a signal indicating sampling is complete.—"LT1016 Ultra Fast Precision Comparator," data sheet, Linear Technology Corporation.

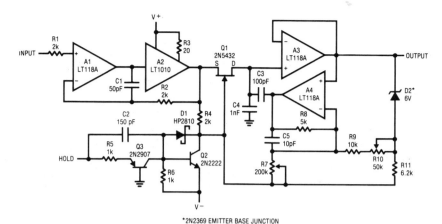

*2N2369 EMITTER BASE JUNCTION

Track and Hold Circuit. 5-MHz track and hold circuit has a 400-kHz power bandwidth driving ± 10 V. A buffered-input follower drives the hold capacitor, C4, through Q1, a low resistance FET switch. The positive-hold command is supplied by TTL logic, with Q3 level shifting to the switch driver Q2. The output is buffered by op amp A3. This circuit is equally useful as a fast acquisition sample and hold.—"LT1010 Fast ± 150 mA Power Buffer," data sheet, Linear Technology Corporation.

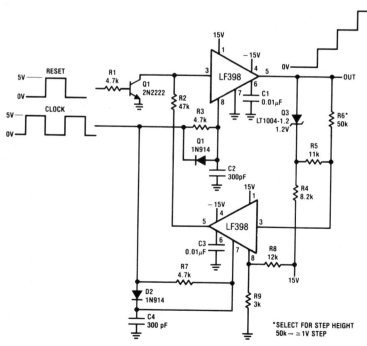

Staircase Generator. Two LF398 sample and hold amplifiers are used to produce an output which increases in discrete steps. Resistor R6, 50 K in the circuit shown, selects the step height. 50 K gives a step height of approximately 1 V. Clock and reset signals control the output rate.—"LF198A/ LF398A/LF198/LF398 Precision Sample and Hold Amplifier," data sheet, Linear Technology Corporation.

*SELECT FOR STEP HEIGHT
50k → ≈1V STEP

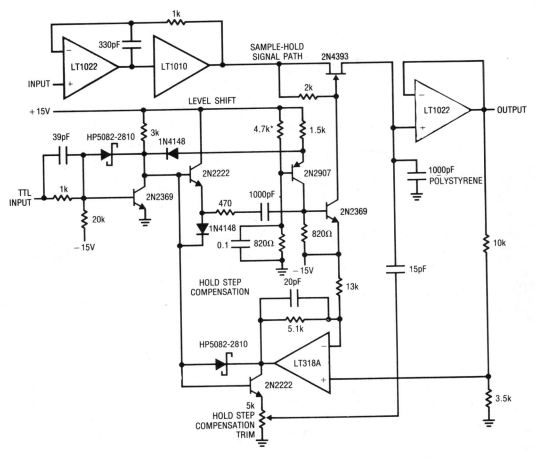

Fast, Precision Sample and Hold. LT1022 devices can be either LT1022 or LT1056 op amps. Circuit has a 16-ns aperture time and a 1-μs acquisition time to 0.01%.—"LT1022 High Speed Precision JFET Input Operational Amplifier," data sheet, Linear Technology Corporation.

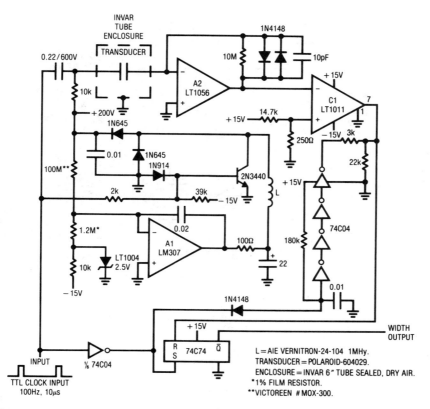

Acoustic Thermometer.   Circuit measures temperature by utilizing the relationship between the speed of sound and temperature in a medium. The inherent time-domain operation of this circuit allows a direct conversion into a digital output A1. The inductor and its associated components form a simple, flyback-type, regulated 200-V supply which biases the transducer. The transducer is a Polaroid 604029 element mounted at one end of a sealed, 6-in. long Invar tube.—"Some Techniques for Direct Digitization of Transducer Outputs," Linear Technology Corporation.

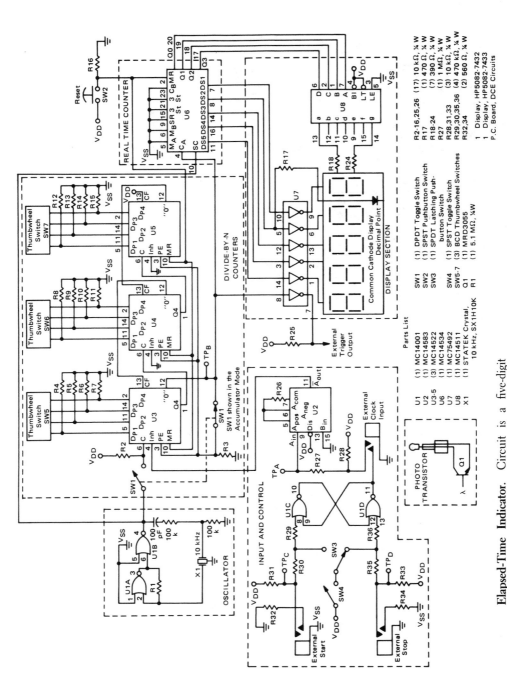

Elapsed-Time Indicator. Circuit is a five-digit elapsed-time indicator with a maximum event of 9999.9 seconds and a five-digit accumulator with a maximum counting capability of 99,999 counts multiplied by N, where N is a prescaler number for 1–999. Heart of the circuit is MC14534 CMOS five-decade counter IC. Oscillator circuit uses a 10-kHz STATEK crystal with a calibration accuracy of ± 0.01%. Original applications note includes construction details.—"Five Digit Accumulator/Elapsed Time Indicator," AN-743, Motorola, Inc.

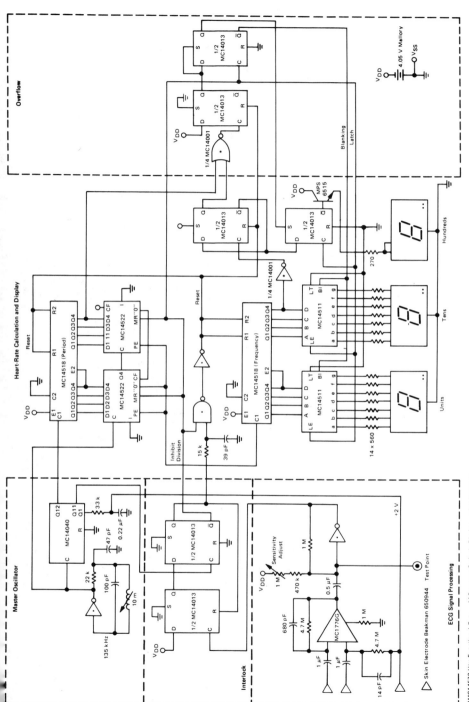

Heart Rate Monitor. Micropower MC1776 operational amplifier and CMOS digital ICs are used in a heart rate monitor with digital readout for each beat. Accuracy is better than 5% under "real world" conditions. Input signals are obtained from electrodes attached to the sternum. Power source is a triple mercury-cell providing 4.05 V at 1 ampere-hour. Power consumption is 600 μW, which allows the battery to last approximately a year.—"A Personalized Heart Rate Monitor with Digital Readout," AN-714, Motorola, Inc.

101

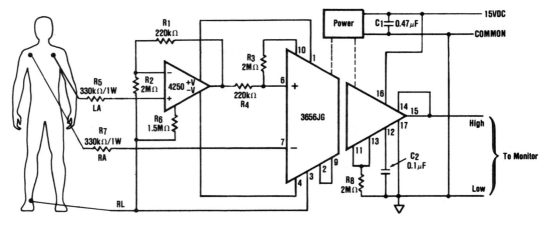

Electrocardiograph Amplifier. 3656 op amp has very high isolation to withstand inadvertent applications of defibrillation pulses while the patient is being monitored. Heart pulses are accurately amplified with a frequency response of DC to 3 kHz.—"Design and Application of Transformer-Coupled Hybrid Isolation Amplifier Model 3656," Applications Note, Burr-Brown.

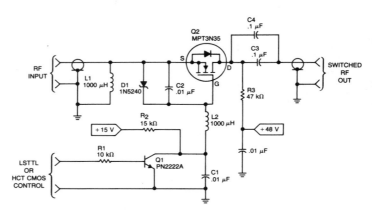

RF Power Switch. Operates at 1.7 MHz with 50-ohm input and output impedances. Its "on" loss is 0.2 dB and its "off" isolation is 30 dB. Circuit requires 1 mA from a +15 V source when "on."—R. Rouquette, "RF Power Switch," TMOS *Power FET Design Ideas*, p. 48, Motorola, Inc.

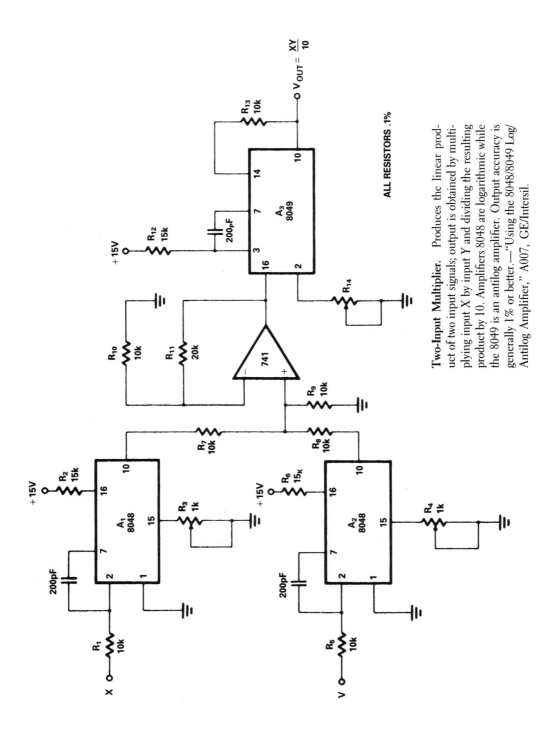

$$V_{OUT} = \frac{XY}{10}$$

ALL RESISTORS .1%

Two-Input Multiplier. Produces the linear product of two input signals; output is obtained by multiplying input X by input Y and dividing the resulting product by 10. Amplifiers 8048 are logarithmic while the 8049 is an antilog amplifier. Output accuracy is generally 1% or better.—"Using the 8048/8049 Log/Antilog Amplifier," A007, GE/Intersil.

103

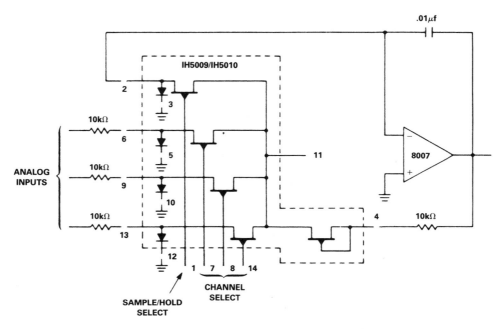

Three-Channel Multiplexer with Sample and Hold.
IH5009/IH5010 CMOS analog switch is used with a
8007 sample and hold amplifier. The input to be
sampled is selected by applying a control voltage to

pin 7, 8, or 14 while the sample and hold function is
controlled by a voltage to pin 1. No external power
source for the IH5009/5010 is required.—"The
IH5009 Analog Switch Series," A004, GE/Intersil.

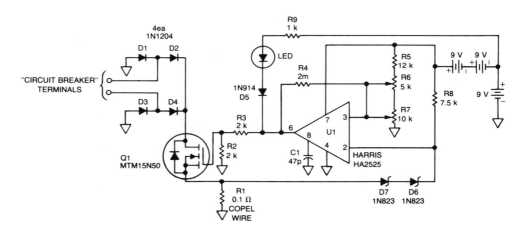

Electronic Circuit Breaker.   This 115-V AC electronic circuit breaker uses the low drive power (low
on resistance and with a fast turnoff) of the TMOS
MTM15N50. Trip point is adjustable, LED fault-
indication is provided, and battery power gives com-

plete circuit isolation. Circuit response time is close
to that of the FET itself.—D. Newton, "High Speed
Electronic Circuit Breaker," *TMOS Power FET Design Ideas*, p. 17, Motorola, Inc.

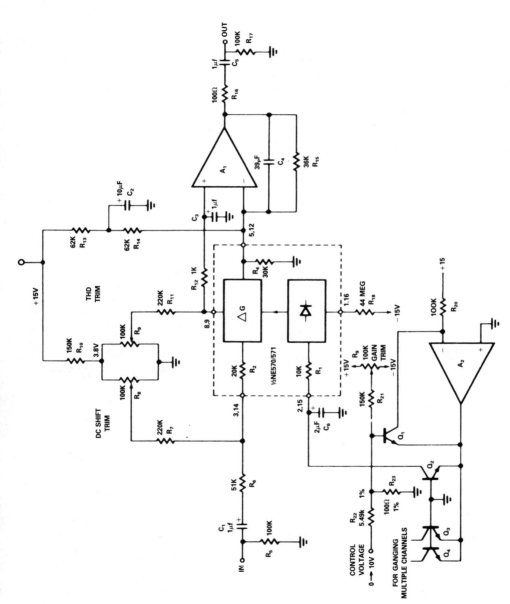

Voltage-Controlled Attenuator. NE570/571 compandor IC serves as major element in a circuit which *reduces* an input signal in response to a control voltage of 0–10 V. Circuit shuts completely off at slightly more than 9 V of control voltage. At lower voltages, attenuation is linear at a rate of −6 dB per V, but steeply increases in a non-linear manner once a total attenuation of −50 dB has been reached.—"Applications of Compandors: NE570/571/SA571," AN174, Signetics Corporation.

105

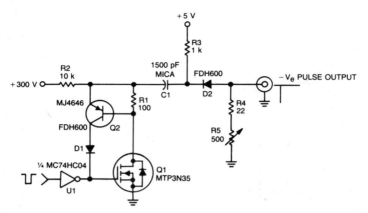

300-V Pulse Generator.
Negative going-pulse is applied to U1, a high speed CMOS buffer, which directly drives the gate of Q1 (MTP3N35). The pulse output across R2 is differentiated by R3/C1 and appears as a negative going-spike at the output terminal.—D. Gray, "300 V Pulse Generator," *TMOS Power FET Design Ideas*, p. 15, Motorola, Inc.

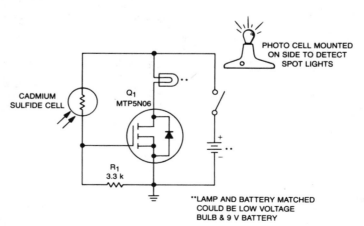

DC Lamp Dimmer.
Potentiometer R3 adjusts lamp brightness. Battery power is stored in C1 for U1, which is a free-running multivibrator whose frequency is determined by R1, R2, R3, R4, and C2. U1 drives the gates of Q1, turning it and the lamp on and off at a rate proportional to the multivibrator duty cycle.—M. Molnar, "DC Lamp Dimmer," *TMOS Power FET Design Ideas*, p. 21, Motorola, Inc.

Light-Controlled Lamp Switch.
When very little light falls on the photocell, its internal resistance is several megohms and R1 keeps the gate of Q1 at nearly 0 V, which keeps it off. When light hits the photocell, its resistance drops to several hundred ohms, raising Q1's gate voltage, turning it on and applying power to the lamp.—D. Baldridge, "Light Controlled Lamp Switch," *TMOS Power FET Design Ideas*, p. 19, Motorola, Inc.

13
Modem Circuits

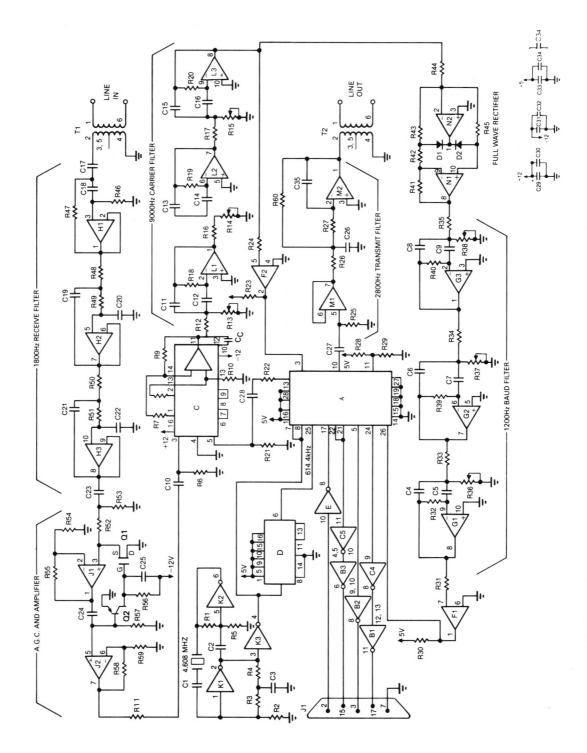

INTEGRATED CIRCUITS		
A	XR-2123	EXAR
B	XR-1488	EXAR
C	XR-2208	EXAR
D	DM-74193	National
E	XR-1489	EXAR
F	LM-339-N	Texas Instruments
G	XR-4741	EXAR
H	XR-4741	EXAR
J	XR-1458	EXAR
K	F-7404	Fairchild
L	XR-4741	EXAR
M	XR-1458	EXAR
N	XR-4741	EXAR

CAPACITORS			
C1	82 pf	C19	.01 μf
C2	.0022 μf	C20	.001 μf
C3	.033 μf	C21	.01 μf
C4	033 μf	C22	100 pf
C5	.033 μf	C23	2.2 μf
C6	.033 μf	C24	2 μf
C7	.033 μf	C25	10 μf
C8	.033 μf	C26	.033 μf
C9	.033 μf	C27	1 μf
C10	.1 μf	C28	.1 μ
C11	.0033 μf	C29	4.7 μf
C12	.0033 μf	C30	.1 μf
C13	.0033 μf	C31	4.7 μf
C14	.0033 μf	C32	.1 μf
C15	.0033 μf	C33	4.7 μf
C16	.0033 μf	C34	.1 μf
C17	.1 μf	C35	.0068 μf
C18	.1 μf		

RESISTORS							
R1	1.2K	R21	2K	R41	10K
R2	2.2K	R22	100K	R42	10K
R3	2.2K	R23	10K	R43	10K
R4	2.2K	R24	10K	R44	10K
R5	2.2K	R25	1M	R45	10K
R6	2K	R26	3.32K	R46	5.76K
R7	24K	R27	2.2K	R47	2.74K
R8	24K	R28	1K	R48	2.61K
R9	50K	R29	1K	R49	75K
R10	50K	R30	10K	R50	7.87K
R11	200K	R31	10K	R51	249K
R12	43.2K	R32	82.2K	R52	120K
R13	1K POT	R33	29.1K	R53	10K
R14	1K POT	R34	29.1K	R54	1K
R15	1K POT	R35	29.1K	R55	68K
R16	43.2K	R36	500Ω POT	R56	1M
R17	43.2K	R37	500Ω POT	R57	10K
R18	109K	R38	500Ω POT	R58	4.7K
R19	109K	R39	82.2K	R59	1K
R20	109K	R40	82.2K	R60	6.8K

TRANSISTORS			
Q1	2N4861	Q2	2N4403

TRANSFORMERS		DIODES	
T1	T2220	D1	IN914
T2	T2220	D2	IN914

CONNECTOR	
J1	RS232

Bell 201/CCITT Standard V.26, 2400-BPS Modem. XR-2123 PSK modulator/demodulator IC serves as key with additional analog and digital circuitry. The digital scrambler/descrambler section is a pseudo-random pattern generator. A counter circuit provides the baud-clock function. The analog circuitry consists of bit-carrier recovery and baud-carrier recovery sections. Filtering removes out of band signals before application to an AGC function during bit-carrier recovery. Baud recovery is done in a similar manner. Accompanying table gives values for parts shown in the schematic.—"XR-2123/XR-2123A PSK Modulator/Demodulator," data sheet, Exar Corporation.

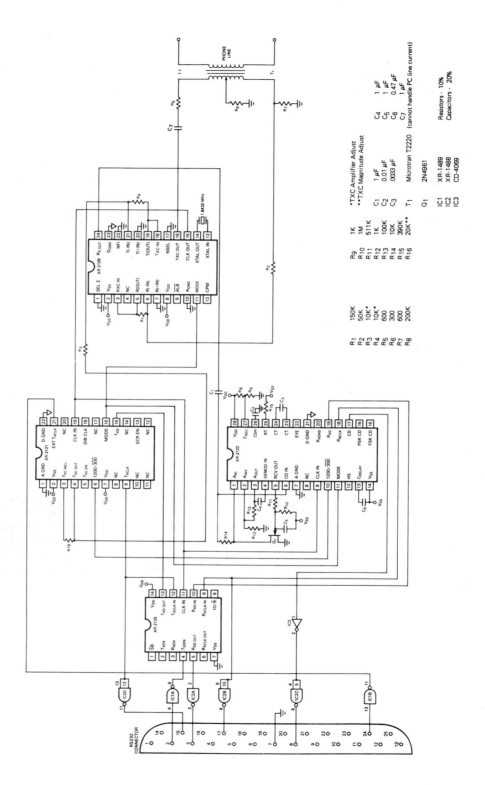

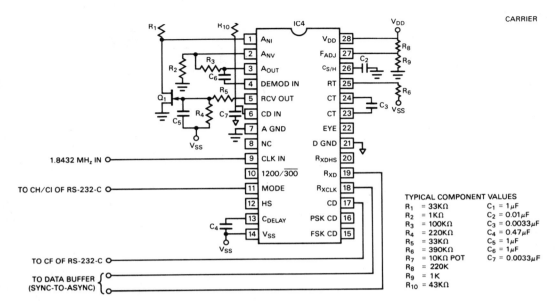

Bell Type 212A Demodulator. Complete demodulator section of a Bell 212A type modem is based on a XR-2122 IC. Circuit is capable of both 1200-bps DPSK and 300-bps FSK demodulation. In 1200-bps operation, an internal 17-bit descrambler provides both descrambled and scrambled output. Speed selection is automatic through an on-chip "handshaking" function.—"XR-2122 Bell 212A Type Demodulator," data sheet, Exar Corporation.

TYPICAL COMPONENT VALUES

R_1	= 33KΩ	C_1	= 1μF
R_2	= 1KΩ	C_2	= 0.01μF
R_3	= 100KΩ	C_3	= 0.0033μF
R_4	= 220KΩ	C_4	= 0.47μF
R_5	= 33KΩ	C_5	= 1μF
R_6	= 390KΩ	C_6	= 1μF
R_7	= 10KΩ POT	C_7	= 0.0033μF
R_8	= 220K		
R_9	= 1K		
R_{10}	= 43KΩ		

◀ **Bell 212A, 1200/300-BPS Modem.** Full duplex modem using the Bell 212A standard is implemented using a XR-2121 modulator IC, a XR-2122 demodulator IC, a XR-2125 buffer device, and a XR-2129 switched-capacitor modem filter. The XR-2129 includes an on-chip clock oscillator, using the 1.8432-MHz crystal shown in the schematic, which is for internal use as well as providing a 1.8432-MHz buffered output.—"XR-2126/7/8/9 Bell 212A/ CCITT V.22 Modem Filters," data sheet, Exar Corporation.

14

Morse Code Circuits

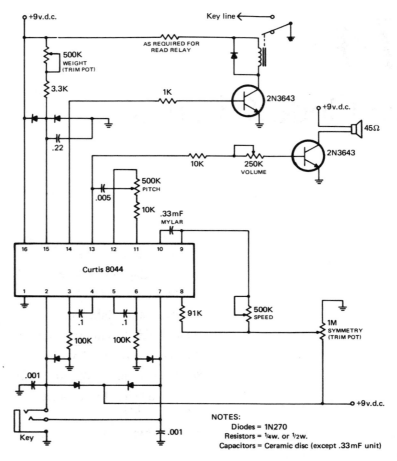

Electronic Keyer.   Designed around a Curtis 8044 keyer IC, circuit provides full iambic keying action for CW transmitters. Keying speed as well as sidetone volume and pitch is continuously adjustable by the potentiometers indicated.—J. Schultz, "Building Ideas for a Consolidated Control Console," *CQ*, July 1986, pp. 11–21.

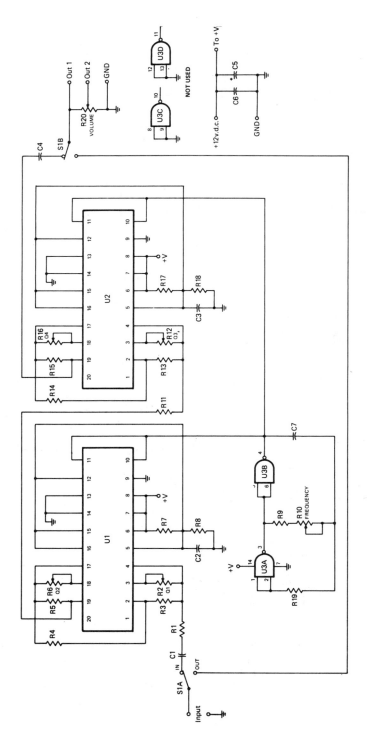

Switched-Capacitor CW Filter. Based on National Semiconductor MF10 universal, monolithic, dual-switched capacitor filter, circuit gives a −3-dB bandwidth adjustable down to 2 Hz and a Q of up to 150. Signals 50 Hz from the center frequency are reduced 40 dB. Potentiometer R10 allows tuning over the range of 740–2700 Hz. C1 and C4 are 0.1 μF, C2, C3, and C5 are 10 μF, tantalum. C6 is 0.01 μF and C7 is 180 pF, mica. R1, R4, R11, R14, and R19 are all 100 K. R2, R6, R12, and R16 are all 100-K linear-taper potentiometer. R3, R5, R13, and R15 are all 180 K. R7, R8, R17, and R18 are all 2.7 K. R9 is 1.8 K. R10 and R20 are both 50 K linear-taper potentiometers. U1 and U2 are the two halves of the MF10, and U3 is a 4011 quad NAND IC.—B. Ives, "How to Build a Switched Capacitor Bandpass CW filter," *CQ*, January 1986, pp. 44–48.

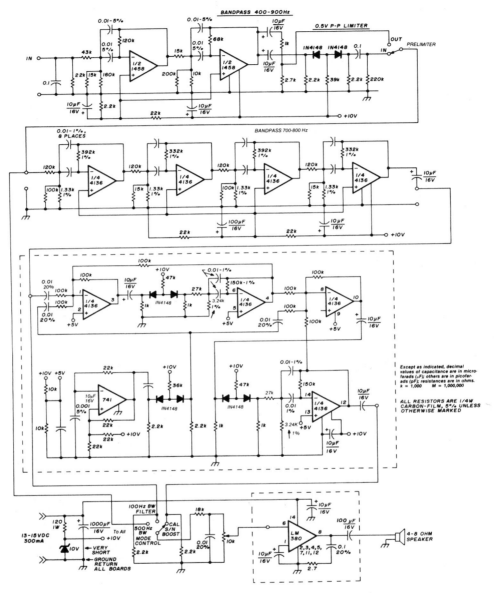

CW Processor.   Based upon "carrier-activated limiter" technique, in which a wideband limiter is constantly switched into upper and lower saturation by a high frequency square wave carrier. Output from the limiter is then fed to a filter that passes a desired signal frequency and removes the square wave carrier.

Technique allows an effective CW bandwidth of 100 Hz to be used without loss of signal and reception of CW signals impossible to hear with ordinary bandwidth filters.—D. Hildreth, "Advanced CW Processor," *ham radio,* December 1986, pp. 25–29.

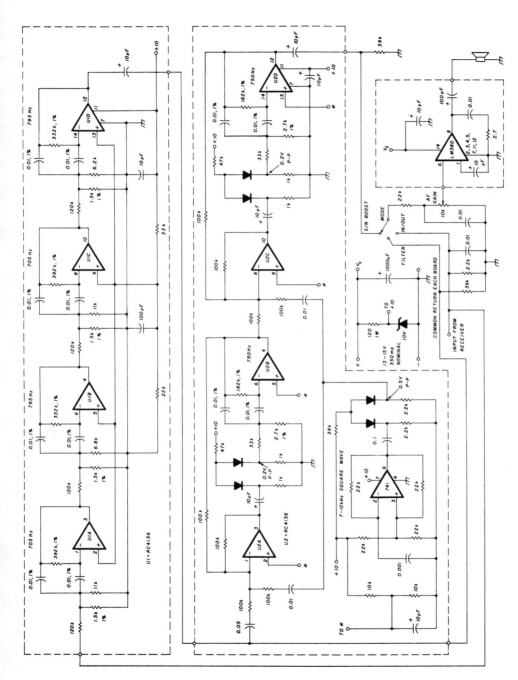

CW Reception Enhancer. Improves signal-to-noise ratio of CW reception by up to 14 db through a carrier-activated limiter preceded and followed by filters. Nominal input voltage is 0.25–0.5 V peak-to-peak at 700–800 Hz. Output can drive a speaker of 4–8 ohms. All diodes are 1N4148 or equivalent.—D. Hildreth, "A Carrier-Activated CW Reception Limiter," *ham radio*, September 1985, pp. 113–120.

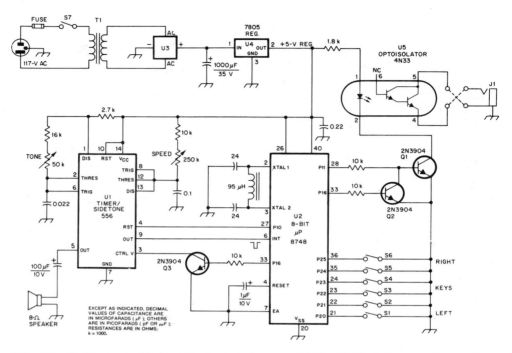

"One-Hand" Electronic Keyer. Closing switches S1 through S6 in proper sequence produces 46 characters representing the alphabet, single digits, and common punctuation in the international Morse code. Circuit is based upon the Intel 8748 microcomputer with internal EPROM; original article includes a program-listing for generation of the Morse code. The 556 dual timer provides a code monitor as well as a speed clock. U3 is a 1-A, 50-PIV diode bridge. The secondary of T1 should provide 12 V at 300 mA. Original article describes key-pressing sequences to produce Morse code characters.—W. Quay and R. Turrin, "Send Error Free Code with One Hand," *QST*, January 1986, pp. 25–28.

Switched-Capacitor CW Filter.   Uses a monolithic RF-5614 switched-capacitor bandpass filter IC to give a bandpass of 300 Hz at -3 and -40 dB. The RF-5614 contains a six-pole Chebyshev class-II bandpass filter whose center frequency is tunable by an external-clock trigger frequency. The external-clock frequency is supplied by an XR-2207 voltage-controlled oscillator (the unlabeled 14-pin IC below the RF-5614A). An alternate method of providing the clock trigger using a XR-2209 precision oscillator IC is shown below. The clock frequency must be equal to 54 times the center frequency desired for the filter; a 1-kHz center frequency will require a 54-kHz clock trigger signal. Circuit using the RF-5614 and XR-2207 will have a center frequency of 1.003 kHz. By substituting the 1-K potentiometer and 1.5-K resistor at pin 4 of the XR-2209, a variable-center frequency of 790–1320 Hz may be obtained.—C. Steer, "Narrow-Minded Filtering," 73 *Magazine*, December 1986, pp. 48–49.

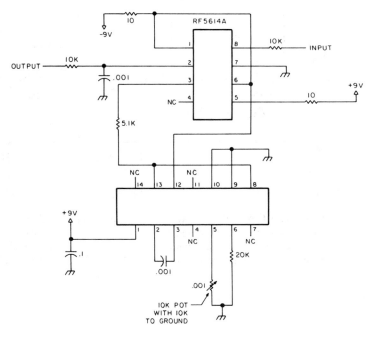

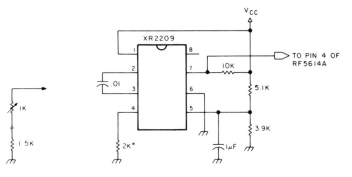

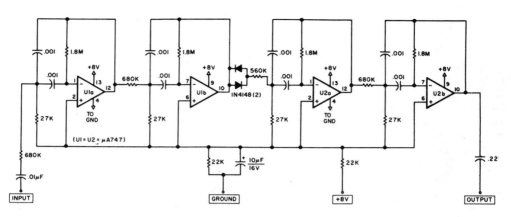

80-Hz Bandwidth CW Filter. Two 747 dual op amp ICs are heart of circuit providing a bandwidth of 80 Hz centered on 750 Hz. Capacitors should be temperature-stabilized and have 5% tolerance or bet- ter. A slight variation in value of the four 27-K resistors may better "match" each of the four op amp stages to each other.—J. Rehak, "Razor Sharp CW," 73 *Magazine*, February 1986, pp. 10–12.

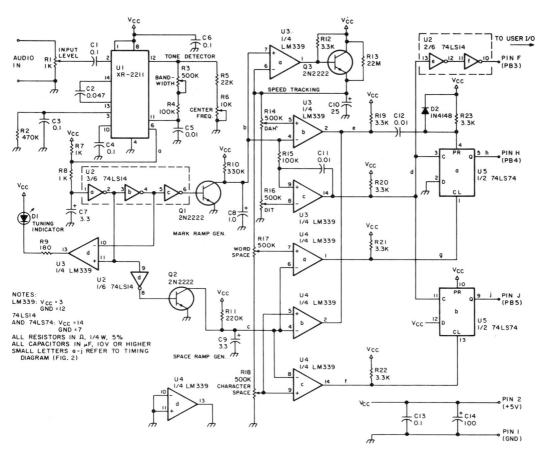

Morse Code to Microcomputer Interface.
Designed for use with microcomputers using the appropriate software to allow conversion of received Morse code characters into characters displayed on a terminal or output to a printer. This circuit accepts Morse code as audio input and makes all decisions regarding speed, length of character elements, and spacing between characters and words. The microcomputer only has to decide which characters have been sent and display or print them. Circuit was designed for use with a Commodore VIC-20 microcomputer but can be easily adapted to other units. Original article includes a short BASIC program listing, written in VIC-20 BASIC, to perform the character recognition-and-display functions; the algorithm it contains can easily be adapted for the versions of BASIC used by other microcomputers. All communication is through the VIC-20 input/output port.—D. Oliver, "The Basics of Computer CW," 73 *Magazine*, April 1986, pp. 59–61.

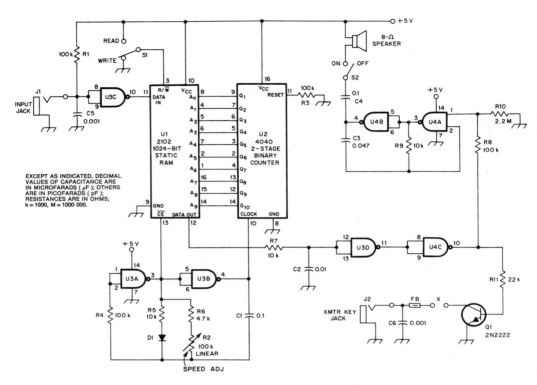

Memory Keyer.   Designed to store and automatically transmit brief CW messages at speeds of up to 150 WPM. Message to be transmitted is entered (using a conventional key) at jack J1 and output to the transmitter is at the point labeled X in the diagram which connects to the transmitter's key jack, J2. R2 adjusts the speed at which the stored message is sent. U3 and U4 are 4011 quad NAND gate ICs. Circuit was originally designed for communications using propagation via ionized meteor trails.—K. Willis, "Meteor Scatter—European Style," QST, November 1986, pp. 35–39.

15

Motor-Control Circuits

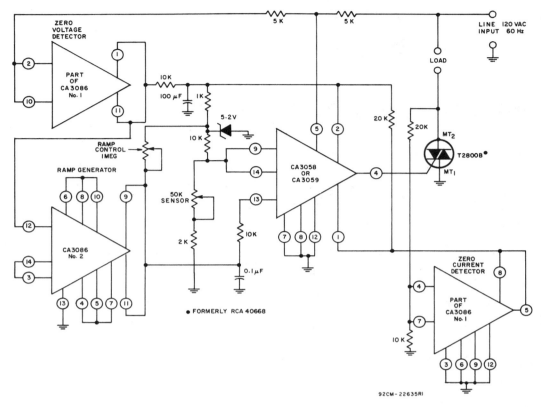

Phase-Control Circuit for AC Induction Motors. A CA3058 or CA3059 zero voltage switch with two CA3086 integrated-transistor arrays form a phase control circuit, which can operate from a line frequency of 50–400 Hz. Circuit can also be used as a light dimmer.—"Features and Applications of RCA Integrated Circuit Zero Voltage Switches (CA3058, CA3059, and CA3079)," ICAN-6182, GE/RCA Solid State.

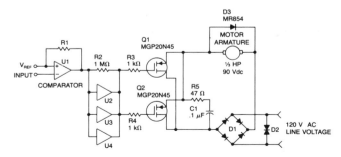

PWM Motor-Speed Control. Two MGP20N45 TMOS devices have their gates pulse width modulated, producing a motor speed proportional to the pulse-width of the incoming digital signal. Input control pulse may be generated by microprocessor or digital logic.—R. Ritzche, "PWM Motor Speed Control," *TMOS Power FET Design Ideas,* p. 32, Motorola, Inc.

Current-Feedback Motor-Control Circuit.   Heart of circuit is a HC2500 linear operational amplifier device. The HC2500 is capable of delivering 7 A of current and a power output up to 100 W RMS. Supply voltage may be up to 75 V.—"HC2500 Multi-Purpose, Low Distortion 7 Ampere Operational Amplifier," File #681, GE/RCA Solid State.

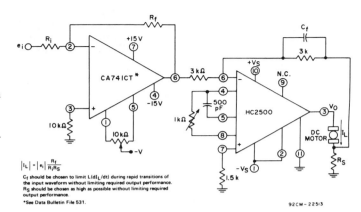

$$|I_L| = |e_i| \frac{R_f}{R_i R_S}$$

C_f should be chosen to limit $L(dI_L/dt)$ during rapid transitions of the input waveform without limiting required output performance. R_S should be chosen as high as possible without limiting required output performance.

*See Data Bulletin File 531.

92CM-22513

Voltage-Feedback Motor-Control Circuit.   Uses HC2500 high power operational amplifier to control speed of a DC motor. Output power of the HC2500 is 100 W at up to 7 A of current; supply voltage may be up to 75 V. Rated bandwidth of the HC2500 is 30 kHz at 60 W.—"HC2500 Multi-Purpose, Low Distortion 7 Ampere Operational Amplifier," File #681, GE/RCA Solid State.

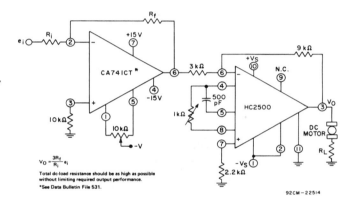

$$V_O = \frac{3R_f}{R_i} e_i$$

Total dc-load resistance should be as high as possible without limiting required output performance.

*See Data Bulletin File 531.

92CM-22514

AC Motor Control.   HC2500 operational amplifier device can supply 100 W RMS of power to control an AC motor. Output current of HC2500 is 7 A and supply voltage can range up to 75 V.—"HC2500 Multi-Purpose, Low Distortion 7 Ampere Operational Amplifier," File #681, GE/RCA Solid State.

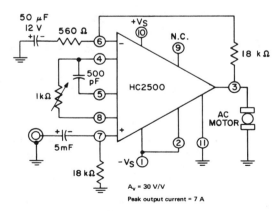

$A_V = 30$ V/V

Peak output current = 7 A

92CS-22515

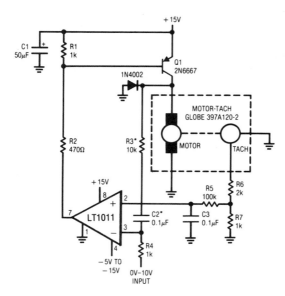

High-Frequency Motor-Speed Controller. Transistor Q1 (2N6667) operates in switching mode. Oscillation frequency of controller is determined by R3 and C2. Control is via a 0–10 V input.—"LT1011/LT1011A Voltage Comparator," data sheet, Linear Technology Corporation.

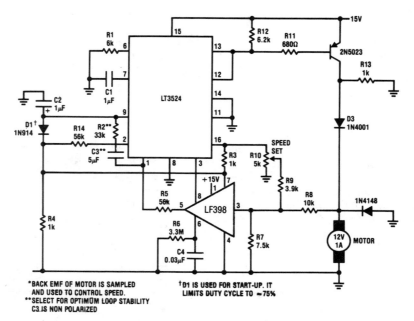

*BACK EMF OF MOTOR IS SAMPLED AND USED TO CONTROL SPEED.
**SELECT FOR OPTIMUM LOOP STABILITY C3 IS NON POLARIZED

†D1 IS USED FOR START-UP. IT LIMITS DUTY CYCLE TO ≈75%

Motor-Speed Controller. LF398 device is a sample and hold amplifier; back EMF of motor is sampled and used to control speed. The value of C3 may have to vary somewhat for maximum circuit stability; it is not a polarized capacitor. D1 is used for start up and limits the duty cycle to 75%.—"LF198A/LF398A/LF198/LF398 Precision Sample and Hold Amplifier," data sheet, Linear Technology Corporation.

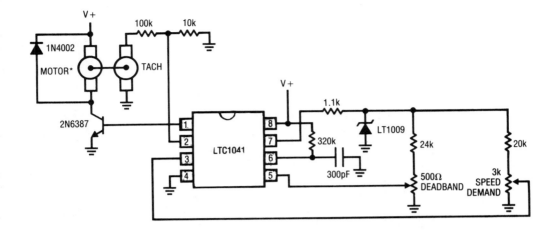

*CANNON CKT26-T5-3SAE

Motor-Speed Controller. Shaded device is a Linear Technology LTC1041 "BANG-BANG" controller IC, which is characterized by turning the control element fully on or fully off to regulate the average value of the parameter to be controlled. The set point input (pin 3) determines the average control-value and the delta input (pin 5) sets the dead band; the dead band is always twice the delta value and is centered around the set-point value.—"LTC1041 BANG-BANG Controller," data sheet, Linear Technology Corporation.

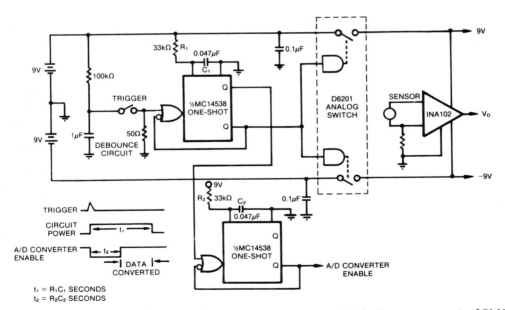

Pulse Controller for Process Monitoring. Circuit pulses on only when it is time to take a measurement in process-control applications, thereby affording significant power savings. The trigger is a momentary-contact switch activated at a set interval (such as 1 second). The two one-shot multivibrators fire when the switch is closed; the first turns on a pair of CMOS switches while the second triggers a DAC to take a reading. After a period determined by the values of R1, C1, R2, and C2, the multivibrators turn off.— "Instrumentation Amp Addresses Power-Miser Circuit Applications," AN-144, Burr-Brown.

127

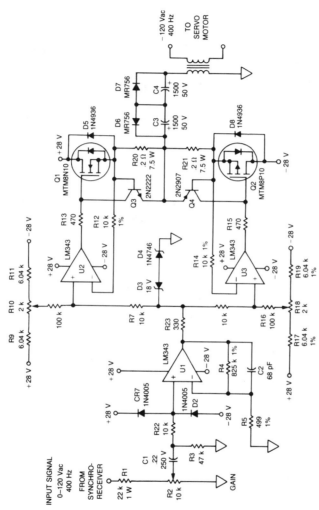

400-Hz Servomotor Amplifier. Signal from a synchro receiver or variable potentiometer is boosted by the LM343 op amp (U1) whose output swing is limited by Zener diodes D3 and D4. The signal is then applied to the other two LM343 op amps, which drives the gates of the MTM8N10 (N-chan-

nel) and MTM8P10 (P-channel) TMOS power FETs. The output transformer steps up the output voltage to −120 V AC for the 400-Hz servomotor. —L. Ducas, "400 Hz Servo Amplifier," *TMOS Power FET Design Ideas,* p. 51, Motorola, Inc.

High Efficiency PWM Speed Control. Main drive-motor is used as a generator/brake to recover battery charge during vehicle braking, increasing the overall range and efficiency of an electric motor. During acceleration, Q1–Q3 receive gate pulses

from U1, and on-line current mode PWM controller IC. As braking occurs, U1 is switched off and the braking IC, U2, and Q2–Q4 are switched on. —L. K. Palmer, "PWM Speed Control and Energy Recovering Brake," *TMOS Power FET Design Ideas,* p. 29, Motorola, Inc.

▶

128

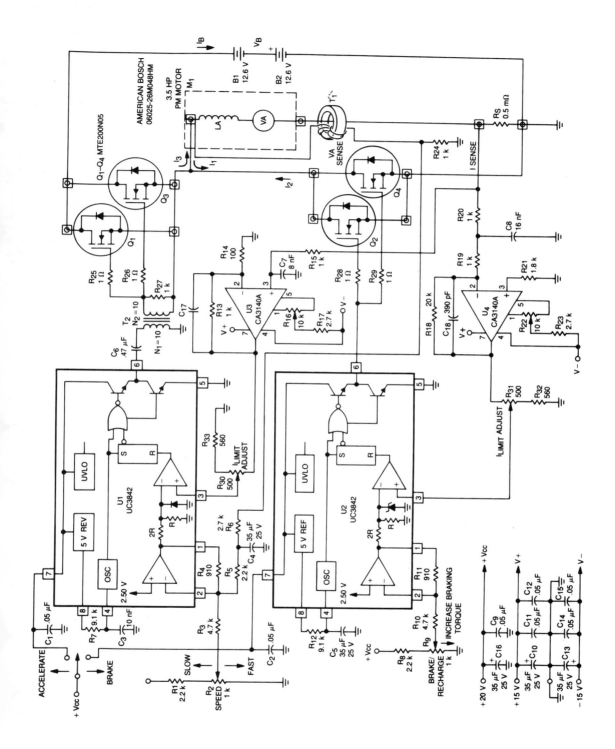

16

Oscillator Circuits

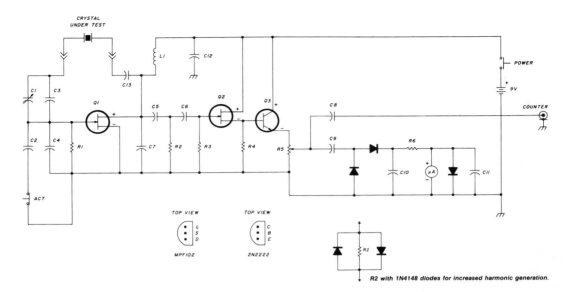

R2 with 1N4148 diodes for increased harmonic generation.

Crystal oscillator parts list.

C1	2.5 – 11 pF
C2	51 pF
C3	56 pF
C4	3 pF
C5	560 pF
C6	100 pF
C7	270 pF
C8	100 pF
C9,C10,C11	0.01 μF
C12	0.47 μF
C13	1500 pF
R1	2.2M ohms
R2	2.2K ohms
R3	1M ohms
R4	4.7K ohms
R5	500 ohms
R6	150 ohms
Q1	MPF 102
Q2	MPF 102
Q3	2N2222

All diodes: IN270
L1 is Hammond Number 1530 C102
10 μH resistance is 1000 ohms (@ 38 mA, maximum).
Meter is 140 μA at 140 microvolts F.S.

100-kHz to 20-MHz Crystal Oscillator. Oscillator Q1 (MPF102) is followed by a buffer (Q2, another MPF102) and an emitter follower with dual output (Q3, 2N2222). The pushbutton switch labeled "ACT" is used for low activity or third-overtone crystals which may need an additional "jolt" to start; most of the time it will be unnecessary. Additional harmonics can be generated by connecting two 1N4148 diodes in parallel back-to-back across a 2.2-K (R2) resistor as shown in the schematic. Accompanying table gives parts used in the schematic.—R. Fransen, "Universal Oscillator Circuit," *ham radio*, April 1986, pp. 38–40.

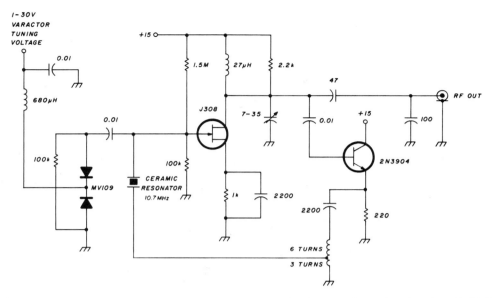

VCO Using Ceramic Resonator. A varactor tuning-voltage of 1–30 V is used to control the output frequency of an oscillator which uses a 10.7-MHz ceramic resonator as its main frequency-determining element. Output can be tuned over a range of approximately 200 kHz based on 10.7 MHz. Input impedance is high due to the FET input amplifier while output impedance is low.—A. Helfrick, "Voltage Controlled Oscillator Uses Ceramic Resonators," *ham radio*, June 1985, pp. 18–26.

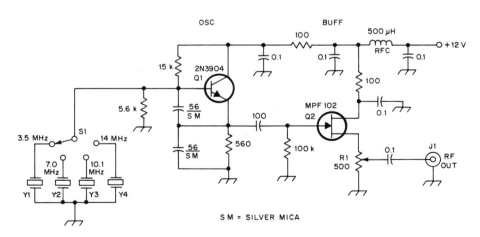

Crystal-Controlled Oscillator. Produces output at frequencies determined by selected crystal. S1 may accommodate additional positions and crystals if desired. The 2N3904 is the oscillator while the MPF102 is a broadband source-follower-buffer. R1, the 500-ohm potentiometer, functions as an attenuator.—D. DeMaw, "Build a Homemade Signal Generator," January 1986, pp. 40–43.

134

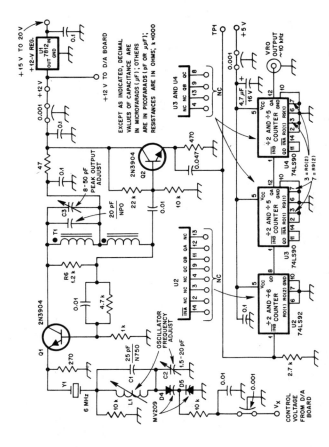

Variable-Reference Oscillator. Improves conventional PLL tuning by providing intermediate frequency steps through changes in the PLL-reference frequency. Circuit was designed for use with a single PLL-based VFO with a tuning range of 1–1.7999 MHz; with such a VFO the original 100-Hz resolution is improved by a factor of 10 with an accuracy on the order of 1 Hz. Circuit was designed in two sections, a D/A converter and linearizer section (*facing page*) and a VXO and frequency-divider section.

Inductor L1 is 35 turns of #32 wire on a ⅜-in. diameter ceramic form with a blue-coded slug. Transformer T1 is wound on a T68-6 toroidal core; the tapped winding is 50 turns of #28 wire tapped 11 turns from one end. Original article gives data on programming the 2716 EPROM and adjustment of the resistors in the D/A and linearizer portions of the circuit.—A Haberstich, "A Variable Reference Oscillator for Synthesized VFOs," *QST*, April 1985, pp. 18–22.

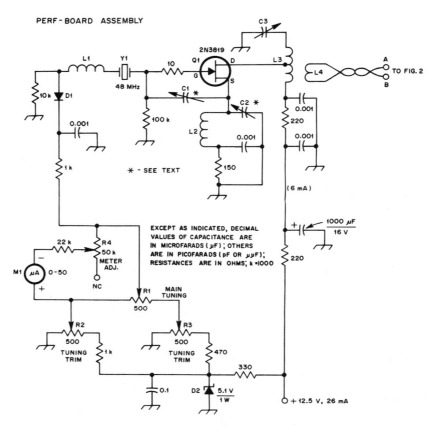

PERF-BOARD ASSEMBLY

145-MHz VXO with 200-mW Output. Constructed in separate oscillator (*above*) and amplifier sections to produce an output suitable for use at 145 MHz or tripled to 435 MHz. Circuit is essentially a VFO synchronized with the third overtone 48-MHz crystal Y1. The free running frequency is determined by the gate- and source-tuned circuits; their relationship establishes the varactor's tuning sensitivity. Stray capacitances can be minimized by mounting the parts on a "perf" board, securing them with push-in terminals, and attaching the perf board to the chassis with 0.75-in. "standoff" connectors. The "varactor" for tuning purposes is actually a 1N4001 (D1) operated in the reverse-bias mode. Inductor L1 should be tailored so that the potential applied to D1 is in the range of 0.3–2 V. L1 is 27 turns of #26 wire wound to a diameter of 5⁄16-in. on a Plexiglas form. L2 is 15 turns of #26 on a 0.25-in. diameter Plexiglas form. L3 is four turns of #14 wire wound to 0.25-in. diameter and 0.50-in. long tapped at two turns. L4 is one turn of #22 stranded wire wound on the end of L3. L5 and L7 are identical to L3; L6 and L8 are identical to L4 and wound on the end of L5 and L7 respectively. C1 through C7 are 5–60 pF trimmer capacitors. RFC1 is 14 turns of #30 wire wound on a 0.5-W, 1-K resistor.—J. Reed, "A Simple 435 MHz Transmitter," *QST*, May 1985, pp. 14–18.

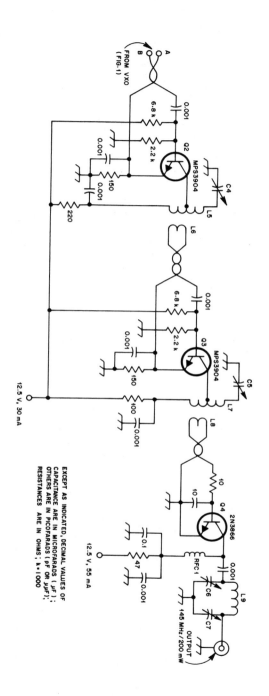

EXCEPT AS INDICATED, DECIMAL VALUES OF
CAPACITANCE ARE IN MICROFARADS (μF);
OTHERS ARE IN PICOFARADS (pF OR μμF);
RESISTANCES ARE IN OHMS; k = 1000

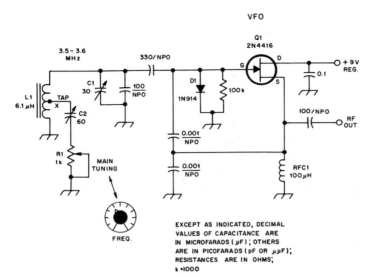

VFO

EXCEPT AS INDICATED, DECIMAL
VALUES OF CAPACITANCE ARE
IN MICROFARADS (μF); OTHERS
ARE IN PICOFARADS (pF OR μμF);
RESISTANCES ARE IN OHMS;
k =1000

3500–3600 kHz VFO with Resistive Tuning. Oscillator uses a 1 K potentiometer (R1) for tuning instead of a variable capacitor. As R1 is adjusted, it forms a series combination with C2, a 60-pF trimmer capacitor. The capacitive reactance formed causes a frequency shift as R1 is adjusted; however, the tuning of R1 is nonlinear. C2 is a "coarse" tuning control that is adjusted once so that R1 covers the desired range.—D. DeMaw, "The Fine Art of Improvisation," *QST*, July 1985, pp. 22–24.

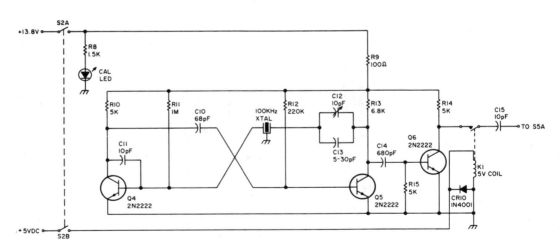

100-kHz Crystal Calibrator. Produces an output rich in harmonics appearing every 100 kHz to past 30 MHz. The 10-pF capacitor, C11, is necessary to guarantee that the oscillator starts each time the power is applied. C12 and C13 may be replaced by one larger variable capacitor if one is available. The case of the crystal should be connected to ground using a short piece of bus wire for added stability. This circuit was originally designed for use with the "Argonaut" amateur radio-transceiver but can be used with other transceivers and receivers.—F. Marcellino, "Get Organized Today with Argo's Helper," *73 Magazine*, March 1986, pp. 44–49.

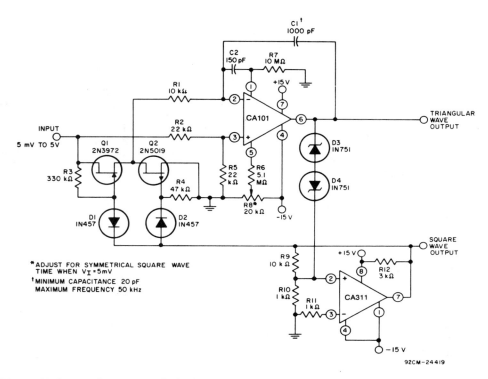

92CM-24419

10 Hz to 10 kHz Voltage-Controlled Oscillator.
Circuit produces an output from 10 Hz to 10 kHz for
a 5 mV to 5 V input. Both triangular and square
wave output signals are available.—"CA311 Voltage
Comparator," File #797, GE/RCA Solid State.

**Selectable Rate, Astable
Multivibrator.**   Switch S1
selects output rate of circuit.
Output rate remains fairly
constant throughout at the rate
selected.—"CA3130 BiMOS
Operational Amplifiers," File
#817, GE/RCA Solid State.

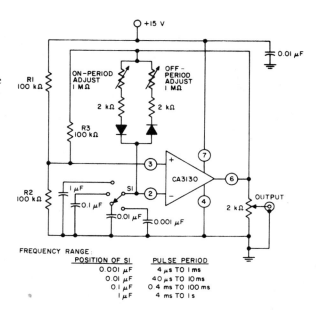

FREQUENCY RANGE:

POSITION OF S1	PULSE PERIOD
0.001 μF	4 μs TO 1 ms
0.01 μF	40 μs TO 10 ms
0.1 μF	0.4 ms TO 100 ms
1 μF	4 ms TO 1 s

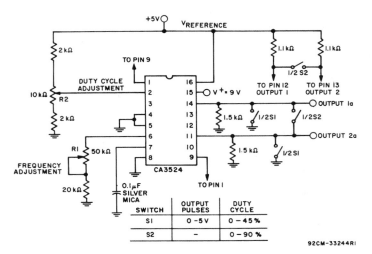

SWITCH	OUTPUT PULSES	DUTY CYCLE
S1	0 -5 V	0 - 45 %
S2	–	0 - 90 %

92CM-33244RI

Low-Frequency Pulse Generator. CA3524 PWM IC is used as low frequency pulse generator. Since all components are on the IC, a regulated 5-V (or 2.5-V) pulse of 0–45% (or 0–90%) on time is possible over a frequency range of 150–500 Hz. Switch S1 is used to go from a 5-V output pulse (S1 closed) to a 2.5-V output pulse (S1 open) with a duty cycle of 0–45%. The output frequency will be roughly half the oscillator frequency when the output transistors are not connected in parallel (75–250 Hz). Switch S2 will allow both output stages to be paralleled for an effective duty cycle of 0–90% with an output frequency range of 150–500 Hz.—"Application of the CA1524 Series Pulse-Width Modulator ICs," ICAN-6915, GE/RCA Solid State.

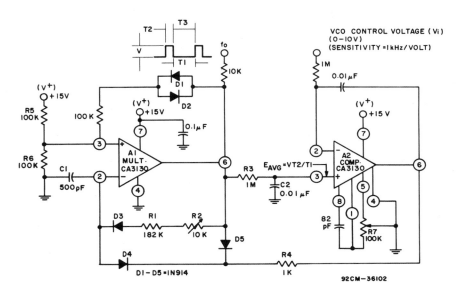

92CM-36102

Precision Voltage-Controlled Oscillator. Operates with a tracking error of 0.05% with excellent linearity up to 10 kHz. Calibration is accomplished by setting V+ to 15 V, V1 (the control voltage) to 10 V, and adjusting R2 until the output is 10 kHz. V1 is then reduced to 10 mV, and R7 is adjusted until the output frequency is 10 Hz.—"Understanding and Using the CA3130, CA3130A, and CA3130B BiMOS Operational Amplifiers," ICAN-6386, GE/RCA Solid State.

100-kHz Crystal-Controlled Oscillator. 100-kHz crystal controls frequency of oscillator based upon CA311 (LM311) voltage comparator IC. Output is TTL compatible.—"CA311 Voltage Comparator," File #797, GE/RCA Solid State.

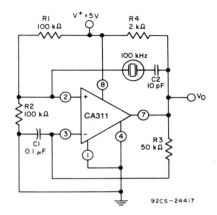

100-kHz Astable Multivibrator. CA311 voltage comparator IC is used in astable multivibrator producing a 100-kHz square wave output. Output is TTL compatible.—"CA311 Voltage Comparator," File #797, GE/RCA Solid State.

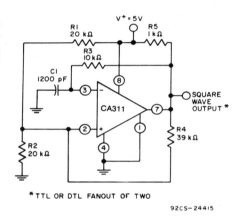

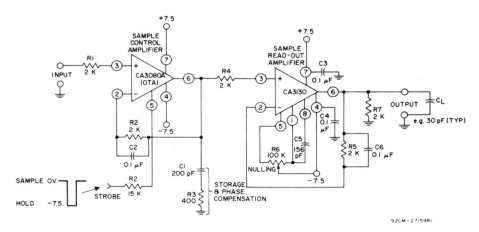

Sample and Hold Circuit. CA3080 transconductance op amp is used in conjunction with CA3130 BiMOS output amplifier. Strobe input voltage is 0 V for sampling and −7.5 V for holding.—"CA3080 Operational Transconductance Amplifiers," File #475, GE/RCA Solid State.

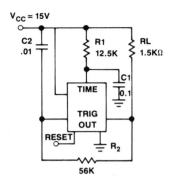

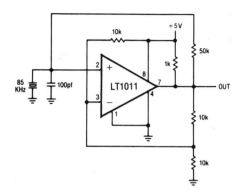

400-Hz Square Wave Oscillator.   Uses one section of a NE558 quad timer device to produce an oscillator for non-precision purposes. The timing interval can be altered by varying C1 and R1; the timing interval is determined by multiplying R1 by C1. Changes in the supply voltage can also affect the output frequency.—"NE558 Applications," AN171, Signetics Corporation.

85-kHz Crystal Oscillator.   LT1011 voltage comparator device serves as main element in oscillator whose frequency is controlled by a 85-kHz crystal.— LT1011/LT1011A Voltage Comparator," data sheet, Linear Technology Corporation.

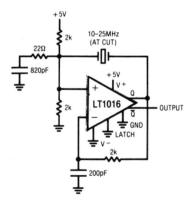

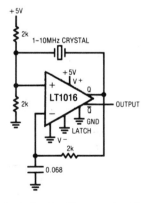

10–25-MHz Crystal Oscillator.   IC is a LT1016 high speed comparator device which permits oscillator to cover the range of 10–25 MHz depending upon the fundamental value of the crystal. Power supply of 5 V permits TTL-compatible output.— "LT1016 Ultra Fast Precision Comparator," data sheet, Linear Technology Corporation.

1–10-MHz Crystal Oscillator.   Oscillation frequency is determined by the fundamental frequency of the crystal. LT1016 high speed comparator device is the main active element.—"LT1016 Ultra Fast Precision Comparator," data sheet, Linear Technology Corporation.

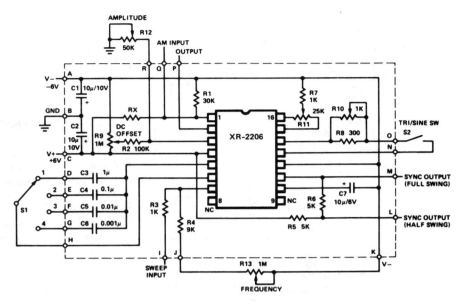

High Quality Function-Generator. Provides sine, triangular, and square wave outputs over a frequency range from 1 Hz to 100 kHz in four ranges as determined by switch S1. Potentiometer R13 varied output frequency over each range. The peak-to-peak values of the sine and triangular outputs can range from 0–6 V. Output frequency can be modulated or swept by applying an external control voltage to terminal label as shown. Output stability is excellent.—"A High Quality Function Generator System Using the XR-2206," AN-14, Exar Corporation.

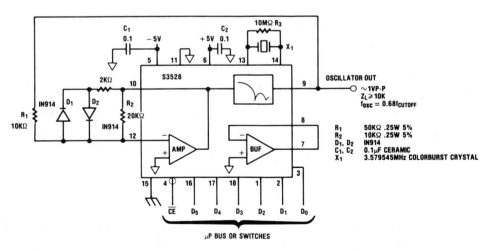

Digitally Tuned Audio Oscillator. S3528 programmable low pass filter IC is configured as an audio oscillator whose output frequency depends upon the digital signals received at pins 1, 2, 3, 4, 16, 17, and 18. Distortion is quite low.—"S3528 Programmable Low Pass Filter," data sheet, Gould/AMI Semiconductors.

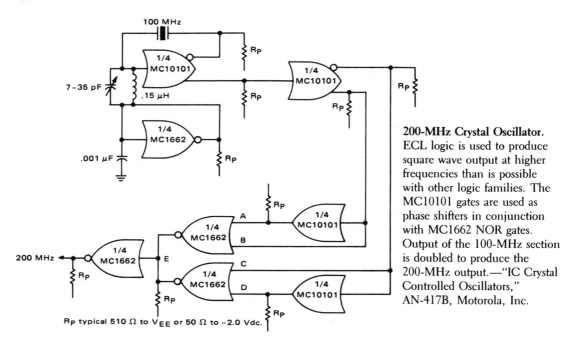

200-MHz Crystal Oscillator. ECL logic is used to produce square wave output at higher frequencies than is possible with other logic families. The MC10101 gates are used as phase shifters in conjunction with MC1662 NOR gates. Output of the 100-MHz section is doubled to produce the 200-MHz output.—"IC Crystal Controlled Oscillators," AN-417B, Motorola, Inc.

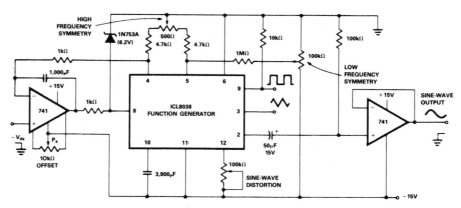

Linear VCO. 741 op amps are used with ICL8038 function generator to produce high linearity between input-sweep voltage and output frequency. Potentiometers allow trimming of output for maximum symmetry and lowest distortion. Square and triangular wave outputs are available from the ICL8038 if needed.—"ICL8038 Precision Waveform Generator/Voltage Controlled Oscillator," data sheet, GE/Intersil.

20 Hz to 20 kHz Variable Audio Oscillator.
ICL8038 precision waveform generator IC produces
sine, triangular, and square waves from 20 Hz to 20
kHz. Output frequency is set by the 10-K potentiometer, while distortion in the output waveform is
minimized by the 100-K potentiometer (it may be
necessary to readjust its setting for the different wave-
forms).—"ICL8038 Precision Waveform Generator/
Voltage Controlled Oscillator," data sheet, GE/
Intersil.

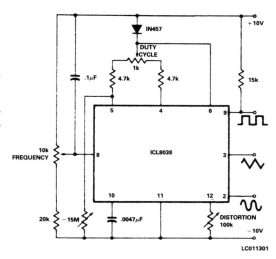

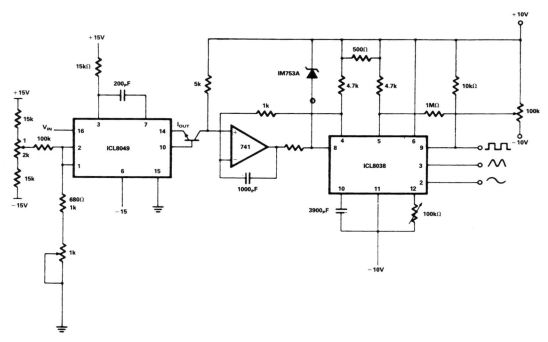

Logarithmic Sweep VCO.   The IC8049 antilog
amplifier controls the frequency output of a ICL8038
waveform generator with the 741 used to linearize
the voltage-frequency response. A common application for this circuit is to use the horizontal sweep of
an oscilloscope as the input for the 8049; this causes
the output of the 8039 to sweep logarithmically
across the audio range.—"Using the 8048/8049 Log/
Antilog Amplifier," A007, GE/Intersil.

17

Phase-Locked Loop Circuits

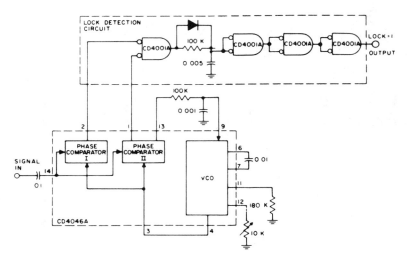

PLL Lock Detection Circuit. Indicates when a PLL circuit or device is in lock. Output is a binary 1 when the PLL is locked and a binary 0 when it isn't. PLL IC used is a 4046 and lock-detection circuit consists of 4001 NOR gates. Circuit shown locks on 10-kHz signals and unlocks on 20-kHz signals.—"The RCA COS/MOS Phase Locked Loop: A Versatile Building Block for Micropower Digital and Analog Applications," ICAN-6101, GE/RCA Solid State.

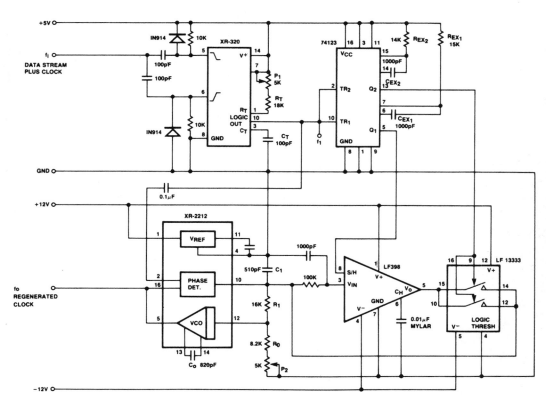

$P_1 =>$ ADJUST SO POSITIVE PORTION OF fi IS EQUAL TO ½ OF THE CLOCK PERIOD
$P_2 =>$ ADJUST FOR 90° PHASE SHIFT BETWEEN f1 and fo WITH fi = f_{CLK}

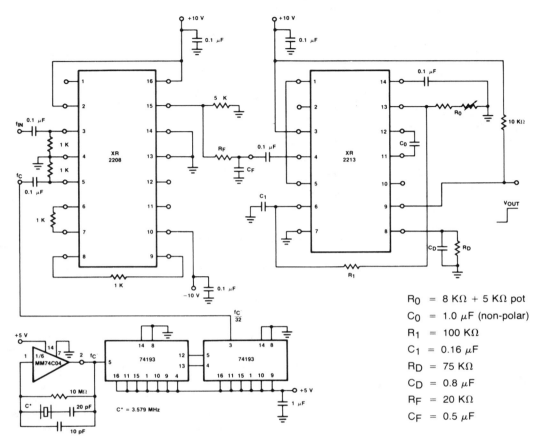

$R_0 = 8\ K\Omega + 5\ K\Omega$ pot
$C_0 = 1.0\ \mu F$ (non-polar)
$R_1 = 100\ K\Omega$
$C_1 = 0.16\ \mu F$
$R_D = 75\ K\Omega$
$C_D = 0.8\ \mu F$
$R_F = 20\ K\Omega$
$C_F = 0.5\ \mu F$

Narrow-Band Tone Decoder. XR-2213 PLL IC is used in conjunction with a XR-2208 analog multiplier as a frequency mixer. It is capable of detecting a 1-Hz tone out of a frequency spectrum of over 1 MHz. It can accept almost any periodic waveform including sine, square, and triangular waves. Operating parameters are determined by values of R0, C0, R1, C1, RD, CD, RF, and CF. In circuit shown, the frequency detection range is 111.7 kHz at ±10 Hz; PLL center frequency is 100 Hz. Original applications note includes full design equations.—"Precision Narrow-Band Tone Detector," AN-21, Exar Corporation.

◄ **Clock Regenerator.** Designed to recover the clock signal embedded in the serial data stream of the common NRZ format (frequently used with floppy disk systems, etc.); the clock signal is necessary to decode the serial data but often cannot be readily extracted using conventional PLL techniques. Circuit shown uses parts values for a clock frequency of 122 kHz and will accept input low levels of 0–0.5 V and high levels from 1.5–5 V; the output is a 10-V peak square wave. The design uses a XR-2212 PLL IC in conjunction with a XR-320 timer as the heart of the system. A 74123 dual monostable multivibrator and 398/13333 are also used for timing and sample and hold purposes.—"Clock Recovery System," AN-19, Exar Corporation.

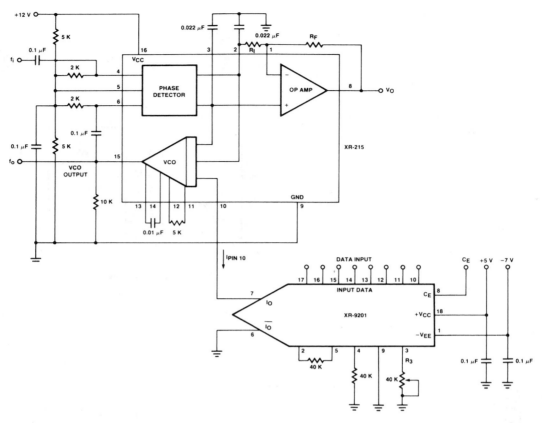

Digitally Programmable PLL. Operation of XR-215 PLL IC is controlled by 8-bit input to XR-9201 DAC device. In design shown, lock range is 5 kHz while capture range is 4 kHz; center frequency is equal to 20 kHz. Original applications note includes full design equations for other center frequencies and lock and capture ranges.—"Digitally Programmable Phase-Locked Loop," AN-24, Exar Corporation.

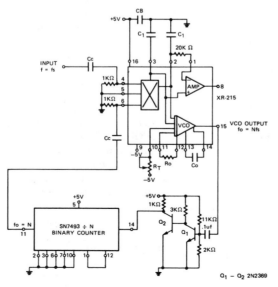

High Frequency Synthesis. System utilizes XR-215 monolithic PLL IC. Output of buffer section is divided down the divider modulus N. When the entire system is synchronized to an input signal at frequency fs, the VCO output (pin 15) is at frequency Nfs, where N is the divider modulus. This permits a large number of discrete frequencies to be synthesized from a given reference frequency.—"High Frequency TLL Compatible Output from the XR-215 Monolithic PLL Circuit," AN-27, Exar Corporation.

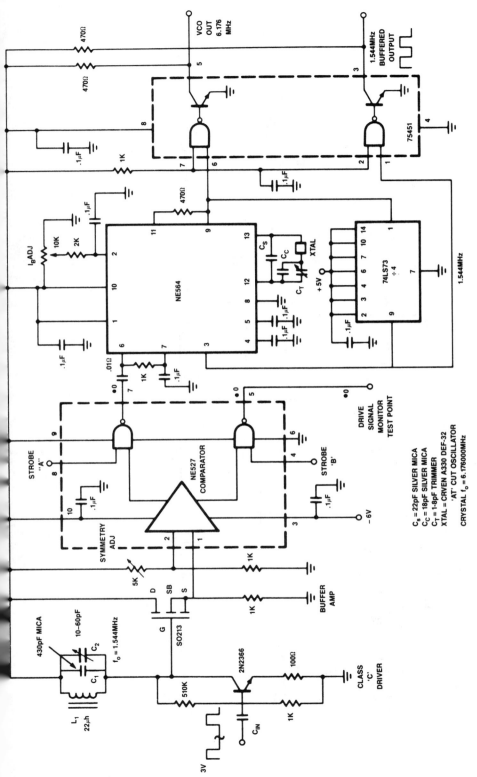

Clock Regenerator. Used to obtain a local clock signal in multiplexed data-transmission systems. A NE564 crystal-controlled PLL is used to reproduce a phase-coherent clock from an asynchronous data stream. The NE564 has TTL-compatible inputs and outputs and can operate at 50 MHz frequencies. In this circuit, it is used as a frequency multiplier incorporating a divide-by-N section in the VCO/phase-detector feedback loop. Crystal used is a fundamental-mode type cut for 6176 kHz.—"Clock Regenerator with Crystal Controlled Phase Locked VCO," AN182, Signetics Corporation.

151

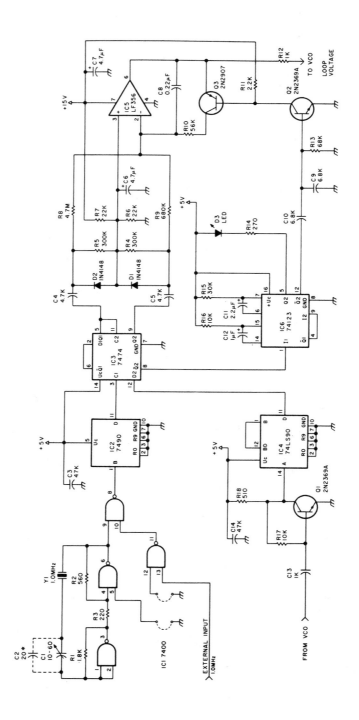

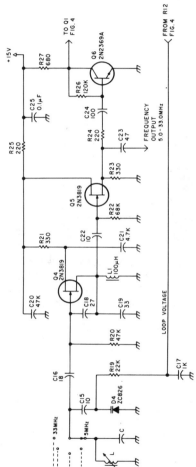

PLL-Based Frequency Synthesizer. Capable of generating frequencies in the range of 5–33 MHz, this circuit consists of a digital section and a VCO. The 1-MHz oscillator section feeds IC2 (a 7490 counter) which operates as a divide-by-5 counter, producing a 200-kHz output. IC3 is a 7474 flip-flop device operating as a frequency detector. At its clock input, pin 11, a 100-kHz signal is supplied. This signal is obtained by dividing 200 kHz by the first flip-flop of IC3. A frequency obtained from the VCO is input at pin 12 of IC3. Before being applied to pin 12, the VCO signal is divided by 10 by IC4, a 74LS90 counter. IC5, the LF356, acts as an integrator. IC6, the 74123, functions as a low pass filter. Transistors Q2 and Q3 serve to neutralize the integrator when the VCO is desynchronized and restore synchronization of the VCO loop. Original article includes construction and alignment information.—E. Ociepka, "Rock Solid RF," *73 Magazine,* April 1986, pp. 74–77.

153

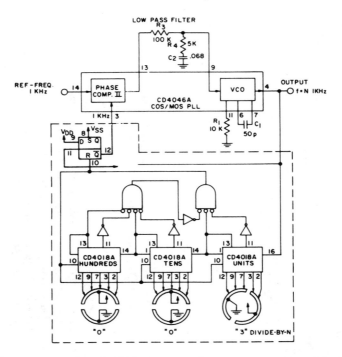

Frequency Synthesizer. Tunes from 3–999 kHz in 1-kHz increments according to the switch positions of the divide-by-N counters. Heart of the circuit is a 4046 PLL IC with a divider circuit, consisting of three 4018 dividers, in the feedback loop of the VCO and comparator sections of the PLL. Reference frequency is 1 kHz.—"The RCA COS/MOS Phase Locked Loop: A Versatile Building Block for Micropower Applications," ICAN-6101, GE/RCA Solid State.

18

Power Supply Circuits

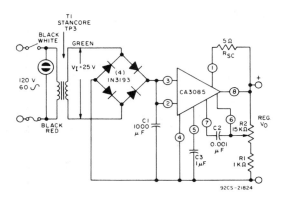

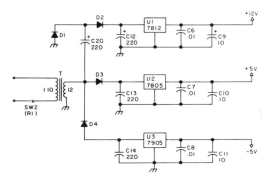

1.8–20 V DC Power Supply. Setting of potentiometer R2 determines output voltage of circuit using CA3085 voltage regulator device.—"Applications of the CA3085 Series Monolithic IC Voltage Regulators," ICAN-6157, GE/RCA Solid State.

Three-Voltage Power Supply. Delivers regulated + 12 V, + 5 V, and − 5 V DC voltages from a single transformer. All diodes are 1-A, 100-V rectifiers.—R. Ketchledge, "The Incredible Inducto-Gauge," 73 *Magazine*, July 1985, pp. 34–36.

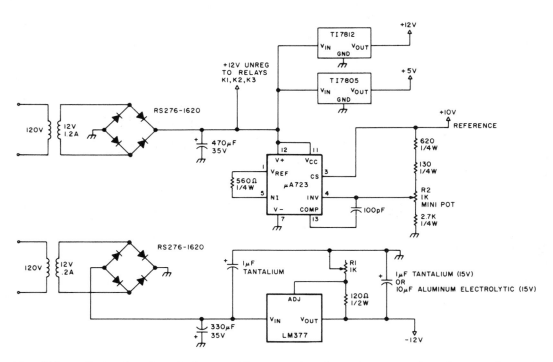

Four-Voltage Power Supply. Produces + 12 V, − 12 V, + 5 V, and + 10 V DC; each may be constructed and used independently of the other. The output of the 12.6-V power-transformer secondary is rectified and filtered to produce approximately + 15 V DC for application to the voltage regulator circuits. All regulators should be mounted on heat sinks.—P. Kaltenbach, "Computer Rotor Control," 73 *Magazine*, June 1986, pp. 64–70.

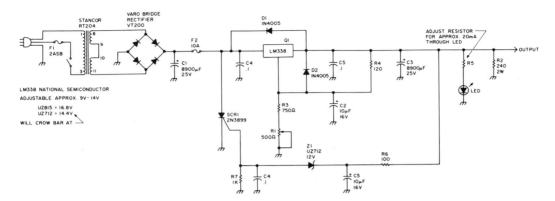

12-V, 5-A Power Supply.   Potentiometer R1 (500 ohms) allows adjustment of the output from 9–14 V. Diodes D1 and D2 prevent C2 and C3 from discharging through the LM338 regulator when the supply is shut off. If the regulator fails, Z1 will turn on at 14.4 V in conjunction with the gate of the SCR, turning on the SCR, and blowing fuse F2. R6 and C5 prevent "false triggering" of this protection.—T. Nolan, "The Peerless Power Pack," 73 *Magazine*, July 1985, p. 38.

Dual Polarity 12-V Power Supply.   Delivers both +12 V and −12 V DC at approximately 1 A.—B. Roehrig, "Tune In the TU-1000," 73 *Magazine*, June 1985, pp. 14–22.

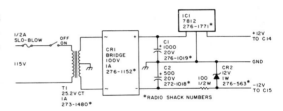

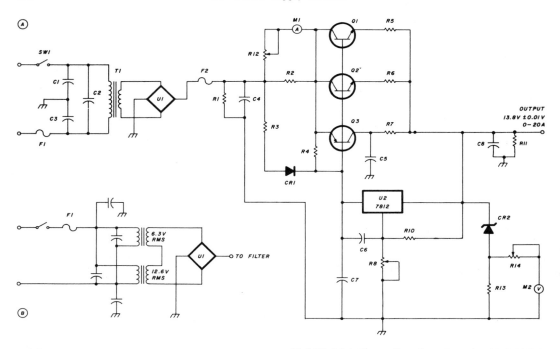

item	description
C1,C2,C3	0.05 µF 600 volt
C4	40,000 µF 35 volt
C5	0.1 to 2.0 µF 50 kilohm (see text)
C6	0.1 µF 50 volt
C7	0.68 µF 50 volt (see text)
C8	0.25 µF 600 volt
CR1	2.5 amp diode 50 volt minimum
CR2	10.0 volt zener, 1.0 watt (see text)
F1	5 amp fuse
F2	25 amp cartridge fuse
M1,M2	0-50 µA DC
Q1,Q2	2N3055 or equivalent
Q3	ECG 129 or TIP 125 or similar PNP
R1	250 ohm 20 watt
R2	0.025 ohm 20 watt
R3	0.5 ohm 2 watt
R4	100 ohm 2 watt
R5,R6	0.1 ohm 5 watt
R7	75 ohm 1 watt
R8	1 kilohm linear
R9	omitted
R10	1.5 kilohm 1/2 watt
R11	150 ohm 5 watt
R12	20 kilohm circuit board potentiometer
R13	100 ohm 1/2 watt
R14	100 kilohm circuit board potentiometer
SW1	toggle switch SPST
T1	120 volt to 18-25 volt, 15-20 amp
U1	25 amp 50 volt rectifier bridge
U2	7812 regulator

13.8-V, 20-A Power Supply. Provides 13.8 V DC at 20 A intermittent and 15 A continuous. Regulation is 0.01 V or better. Inset (labeled *B*) shows details of T1 using transformers in series. Accompanying table gives values of parts shown in schematic.—G. Thurston, "Designing Low Voltage Power Supplies," *ham radio*, March 1985, pp. 46–59.

+8–14 V DC Adjustable-Output Power Supply.
Provides an output from 8–14 V DC at approximately 1 A as determined by setting of the 1-K potentiometer. Circuit was originally designed to be used with a 1296-MHz "down" converter whose frequency was voltage controlled.—C. Houghton, "Build a 1296 Stripper," 73 *Magazine*, October 1985, pp. 40–44.

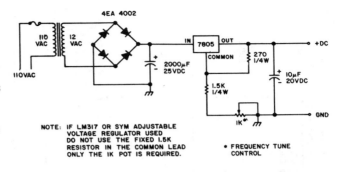

NOTE: IF LM317 OR SYM ADJUSTABLE
VOLTAGE REGULATOR USED
DO NOT USE THE FIXED 1.5K
RESISTOR IN THE COMMON LEAD
ONLY THE 1K POT IS REQUIRED.

• FREQUENCY TUNE
CONTROL

Dual Voltage DC Power Supply.   Provides two positive-polarity DC voltages from approximately 2–18 V. Output voltage is set by 5-K potentiometers connected to the LM317 voltage regulators. Power transformer used has twin 18-V secondaries. Circuit was designed to supply 9-V and 12-V outputs.—J. Schultz, "Building Ideas for a Consolidated Control Console," *CQ*, July 1986, pp. 11–21.

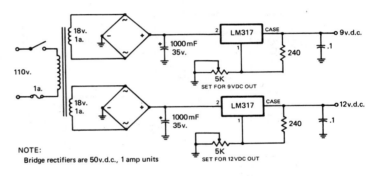

NOTE:
Bridge rectifiers are 50v.d.c., 1 amp units

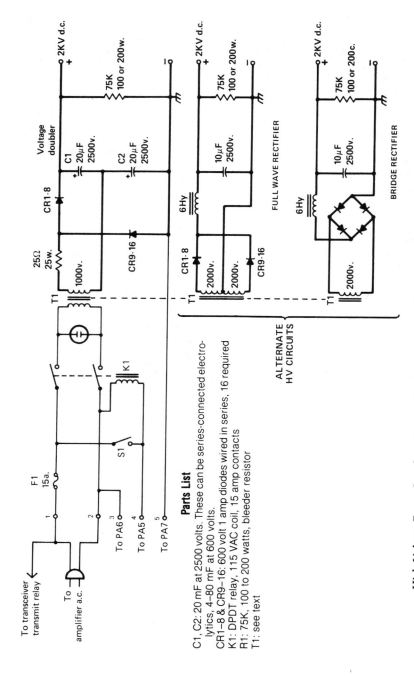

Parts List

C1, C2: 20 mF at 2500 volts. These can be series-connected electro-
lytics, 4–80 mF at 600 volts.
CR1–8 & CR9–16: 600 volt 1 amp diodes wired in series, 16 required
K1: DPDT relay, 115 VAC coil, 15 amp contacts
R1: 75K, 100 to 200 watts, bleeder resistor
T1: see text

High Voltage Power Supply. Designed to power a
1000-W RF amplifier. Transformer used is rated for
980 V at 800 mA. CR1–CR8 and CR9–CR16 are
600-V, 1-A diodes wired in series; 16 in total are required. K1 is a DPDT relay with a 115-V AC coil
and 15-A contacts. C1 and C2 can be 4–80 µF,
600-V electrolytic capacitors connected in series.
Alternate bridge and full wave rectifier configurations
are shown.—I. Wolfe, "A Surplus KW Amplifier,"
CQ, March 1985, pp. 13–19.

Multivoltage Adjustable Power Supply. Circuit offers dual polarity 12-V outputs at 1 A in addition to a variable 1–15 V output at 5 A. T1 has a secondary of 12.6-V RMS at A and T2 has a secondary of 25.6-V RMS at 3 A, center-tapped. CRB1 is a 12 A-bridge rectifier rated at 50 PIV. CRB2 is a 1-A bridge rectifier rated at 50 PIV (1000 PIV rating was used in original design).—J. Carr, "Build a Bench Power Supply," *ham radio*, March 1986, pp. 55–58.

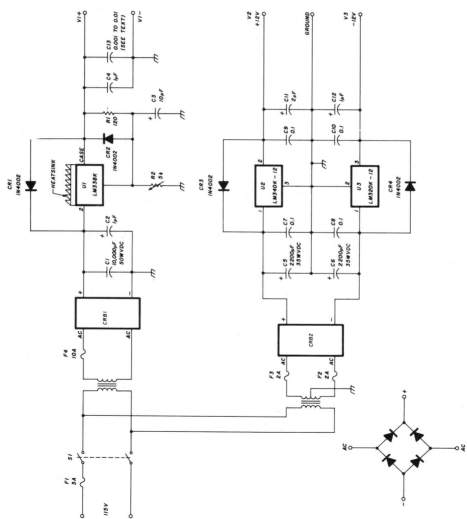

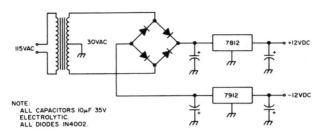

NOTE:
ALL CAPACITORS 10μF 35V
ELECTROLYTIC.
ALL DIODES 1N4002.

Dual Polarity 12-V Power Supply. Circuit provides 12 V DC in positive and negative polarities at approximately 1 A of current. All capacitors are electrolytic, 10 μF rated at 35 V. The four diodes are 1N4002 or equivalent. The power transformer has a 30-V secondary with a center tap to ground.—M. Dooley, "The Commodore Cable Caper," 73 *Magazine*, January 1986, pp. 44–45.

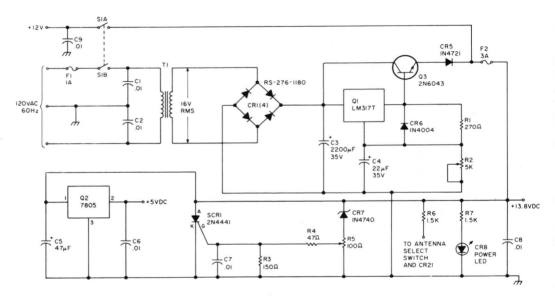

13.8-V DC Power Supply with "Crowbar" Protection Circuit. Allows operation from either a 12-V DC or 120-V AC source as selected by switch S1. R2 adjusts the output of the LM317T voltage regulator device and is adjusted for 13.8-V DC output. R5 sets the control threshold-voltage of the "crowbar" protection circuit; suggested setting is about 15.5 V DC. If the LM317T fails, the gate voltage of SCR1 will reach the "trip" point set by R5. This forces SCR1 into conduction and switches the 13.8-V DC line to ground-potential rapidly, simultaneously exceeding the limits of fuse F2. The 2N6043 pass transistor should be mounted to the chassis using mica washers and thermal grease. Circuit was originally designed for use with "Argonaut" low power amateur radio transceivers but is useful with similar low power devices.—F. Marcellino, "Get Organized Today with Argo's Helper," 73 *Magazine*, March 1986, pp. 44–49.

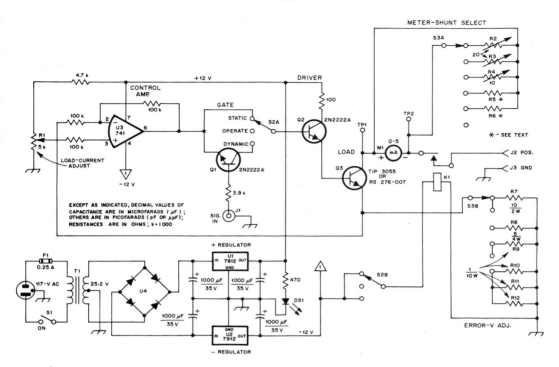

Electronic Load for Power Supplies.   Closely simulates conditions that power supplies will encounter under actual operating conditions and allows such power supplies to be measured and evaluated independently of the circuit(s) they will power. The load presented will be approximately 12 V at current levels from 5 A to 10 m A. This circuit will hold the load current constant during small changes in power supply voltage and allows for both static and dynamic loads for the power supply. R1 should be a 10-turn potentiometer, while R5 and R6 are low resistance shunts made from approximately one meter of insulated #22 wire formed in a coil.—B. Lent, "A Power Supply Performance Tester," *QST*, April 1985, pp. 38–41.

15-V Unregulated Power Supply.   Simple circuit provides 15 V at 1 A. Other desired regulated voltages can be obtained by using voltage divider networks and voltage regulator ICs.—P. Kranz, "Better Ears for the MATVI-40 Transceiver," *QST*, October 1985, pp. 14–20.

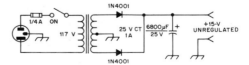

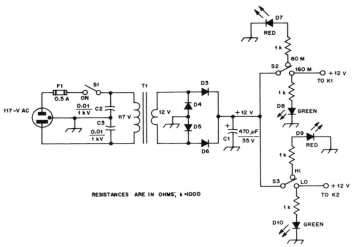

12-V DC Power Supply.
Originally designed to power a
remote-antenna switch, circuit
offers two +12 V DC points
and two indicating LEDs; all
the indicating LEDs and their
current limiting resistors may
be eliminated if not needed for
a specific application. D3–D6
are 1 A, 50-PRV diodes while
D7–D10 are LEDs. T1 should
be capable of supplying 450
mA of current.—D. DeMaw,
"A Remotely Switched,
Inverted-L Antenna," *QST*,
May 1985, pp. 37–39.

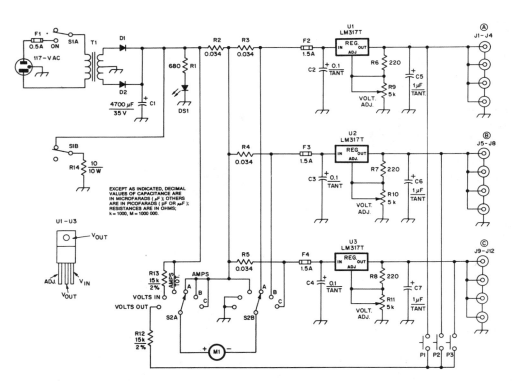

Three-Voltage Power Supply. Provides three
1.2–14-V voltages at approximately 1.5A each. Circuit consists of a single rectifier feeding three separate
regulated power supplies, each using the LM317T
voltage regulator IC. M1 is a 0–1 mA voltmeter with
85-ohm internal resistance. D1 and D2 are 1N5400
or equivalent. T1 has a secondary of 25.2 V, 2 A,
with a center tap.—G. Murphy, "The Super AC
Adapt," *QST*, December 1985, pp. 25–28.

15-V Dual Tracking Regulator.   CA3085 voltage regulator IC is used with a CA3094 operational amplifier which is capable of supplying 100 mA of output current. The positive output voltage is regulated by the CA3085 while the negative output voltage is regulated by the CA3094, which is "slaved" to the regulated positive-voltage supplied by the CA3085.— "Applications of the CA3085 Series Monolithic IC Voltage Regulators," ICAN-6157, GE/ RCA Solid State.

MAX. I_{OUT} = ± 100 mA

REGULATION:

MAX. LINE = $\dfrac{\Delta V_{OUT}}{\left[V_{OUT}\,(\text{INITIAL})\right]\,\Delta V_{IN}}$ x 100 = 0.075 % / V

MAX. LOAD = $\dfrac{\Delta V_{OUT}}{V_{OUT}\,(\text{INITIAL})}$ x 100 = 0.075 % V_{OUT}

(I_L FROM 1 TO 50 mA)

92CM-20560

Absolute-Value Full Wave Rectifier.   During positive peaks, the input signal is fed through the feedback network directly to output. Simultaneously, the positive portion of the input signal drives the output terminal (6) of the inverting amplifier in a negative-going direction so that the 1N914 diode effectively disconnects the amplifier from the signal path. During the negative portion of the input signal, the CA3130 functions as a normal inverting amplifier. —"CA3130 BiMOS Operational Amplifier," File #817, GE/RCA Solid State.

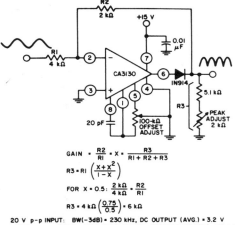

GAIN = $\dfrac{R2}{R1}$ · X = $\dfrac{R3}{R1 + R2 + R3}$

R3 = R1 $\left(\dfrac{X + X^2}{1 - X}\right)$

FOR X = 0.5: $\dfrac{2\,k\Omega}{4\,k\Omega} = \dfrac{R2}{R1}$

R3 = 4 kΩ $\left(\dfrac{0.75}{0.5}\right)$ = 6 kΩ

20 V p-p INPUT: BW(-3dB) = 230 kHz, DC OUTPUT (AVG.) = 3.2 V
1 VOLT p-p INPUT: BW(-3dB) = 130 kHz, DC OUTPUT (AVG.) = 160 mV

92CS-24730

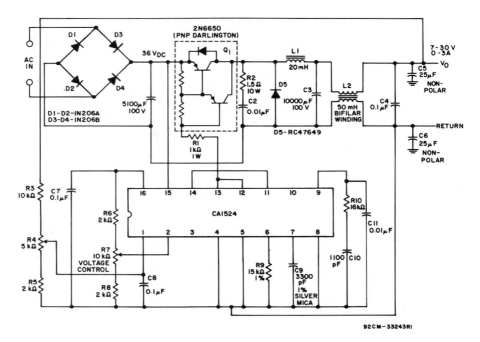

Laboratory Power Supply. Output can range between 7–30 V at up to 3 A; currents as high as 5 A may be drawn with decreased voltage regulation. Heart of circuit is a CA1524 PWM IC. By connecting the two output transistors in parallel, the duty cycle is doubled. Q1 is used as the switching pass element.—"Application of the CA1524 Series Pulse-Width Modulator ICs," ICAN-6915, GE/RCA Solid State.

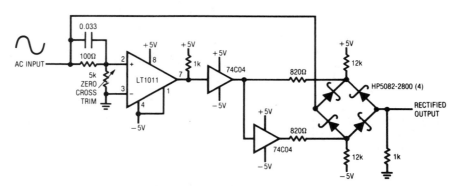

100-kHz Precision Rectifier. Circuit filters 100-kHz AC input into DC. The 5-K potentiometer across pins 2 and 3 of the LT1011 voltage comparator trims the output of the circuit for pure DC.—"LT1011/LT1011A Voltage Comparator," data sheet, Linear Technology Corporation.

5-V Power Supply Monitor. Dual LT1017 low power comparator uses both sections to produce high output if the monitored power supply voltage is between 4.5–5.5 V; output is low if it is not.— "LT1017/LT1018 Micropower Dual Comparator," data sheet, Linear Technology Corporation.

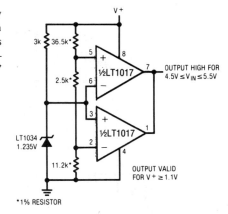

Power Supply Monitor. Flickering of LED gives visual indication of power supply voltage. The flickering rate is 5 Hz at 4.75 V, 3 Hz at 5 V, 1 Hz at 5.25 V, and the LED is off at 6 V. One section of LT1017 comparator is used.—"LT1017/LT1018 Micropower Dual Comparator," data sheet, Linear Technology Corporation.

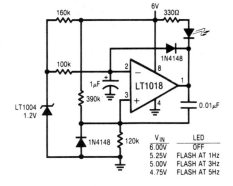

Low Power Switching Regulator. Provides 5-V output from a 9-V source with 80% efficiency and 50-mA output capability. The 100-pF capacitor ensures clean switching, while the 1-μF capacitor helps provide low battery-impedance at high frequencies.—"Power Conditioning Techniques for Batteries," Application Note 8, Linear Technology Corporation.

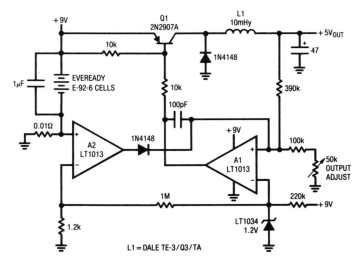

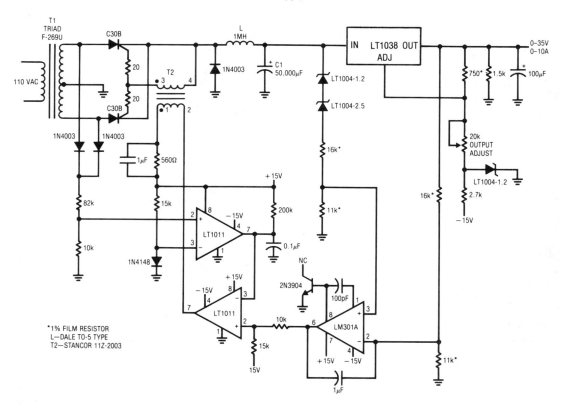

*1% FILM RESISTOR
L—DALE TO-5 TYPE
T2—STANCOR 11Z-2003

Wide-Range Regulated Supply. Output ranges from 0–35 V at 0–10 A. Shaded device is a LT1038 10-A positive, adjustable voltage regulator IC; a LM396 device may also be substituted. SCR preregulation is used to reduce power dissipation. Inductor L is a Dale T0-5 or equivalent, and transformer T2 is a Stancor 11Z-2003 or equivalent. Asterisked resistors are 1% film types.—"LT1038 10 Amp Positive Adjustable Voltage Regulator," data sheet, Linear Technology Corporation.

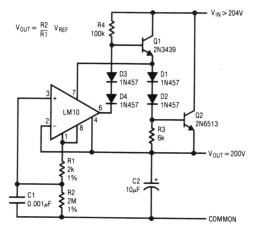

$$V_{OUT} = \frac{R2}{R1} \; V_{REF}$$

High Voltage Regulator. Output voltage of 200 V can be maintained with input voltages in excess of 204 V.—"LM10 Low Power Op Amp and Reference," data sheet, Linear Technology Corporation.

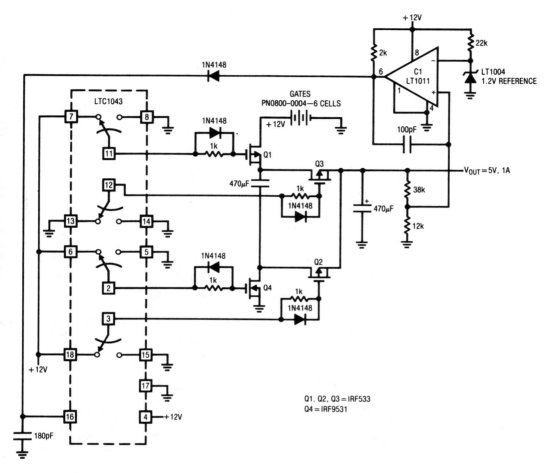

"Inductorless" Switching Regulator. No inductor is used in this regulator which can supply 1 A of output current. The LTC1043 switched capacitor IC provides non-overlapping complementary drive to the power MOSFETs (Q1–Q4). The MOSFETs are arranged so that C1 and C2 are alternately placed in series and then in parallel. During the series phase, the +12-V battery's current flows through both capacitors, charging them and furnishing load current. During the parallel phase, both capacitors deliver current to the load. The circuit constitutes a large scale, switched capacitor, voltage divider which is never allowed to complete a full cycle. The high transient currents are easily handled by the power MOSFETs and overall efficiency is about 83%. —"Power Conditioning Techniques for Batteries," Application Note 8, Linear Technology Corporation.

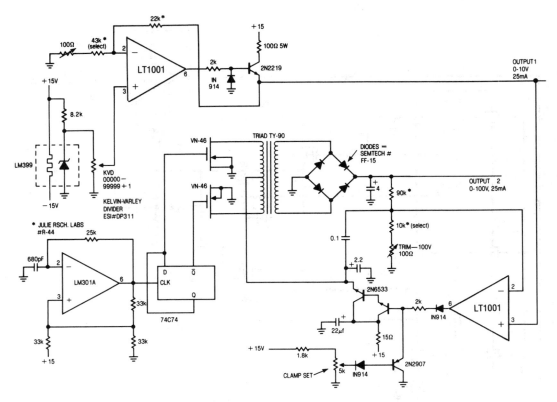

Dual-Output, Precision Power Supply. Provides 0–10 V in 100-μV steps and 0–100 V in 1-mV steps from dual polarity 15-V supply. Shaded devices are LT1001 op amps. Accuracy is excellent through its range.—"LT1001 Precision Operational Amplifier," data sheet, Linear Technology Corporation.

5-V, 50-A Switching Supply. Half bridge design accepts 220 V AC, 50-Hz input and delivers a dual polarity 5-V output at 50 A. Heart of circuit is a MC3420 switchmode regulator control circuit, which provides oscillator, pulse-width modulation, and pulse-routing functions. Load regulation is 0.25% or better with a maximum output ripple of 60 mV peak-to-peak. Efficiency is 75%.—"New ICs Perform Control and Ancillary Functions in High Performace Switching Supplies," EB-78, Motorola, Inc.

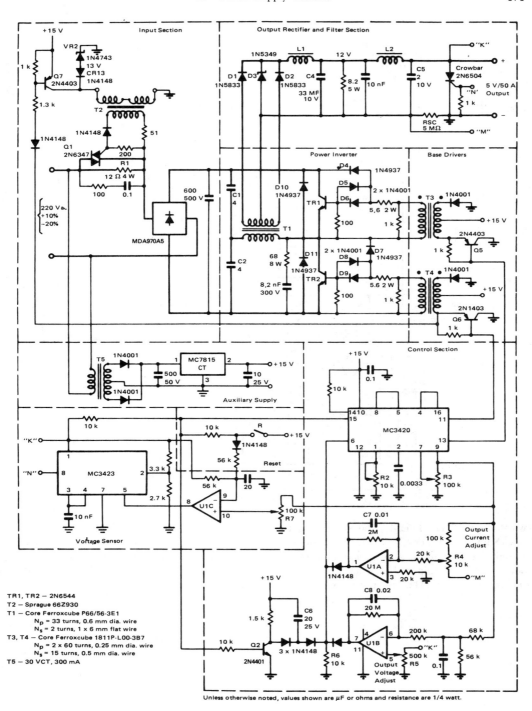

TR1, TR2 — 2N6544
T2 — Sprague 66Z930
T1 — Core Ferroxcube P66/56-3E1
   N_p = 33 turns, 0.6 mm dia. wire
   N_s = 2 turns, 1 x 6 mm flat wire
T3, T4 — Core Ferroxcube 1811P-L00-3B7
   N_p = 2 x 60 turns, 0.25 mm dia. wire
   N_s = 15 turns, 0.5 mm dia. wire
T5 — 30 VCT, 300 mA

Unless otherwise noted, values shown are µF or ohms and resistance are 1/4 watt.

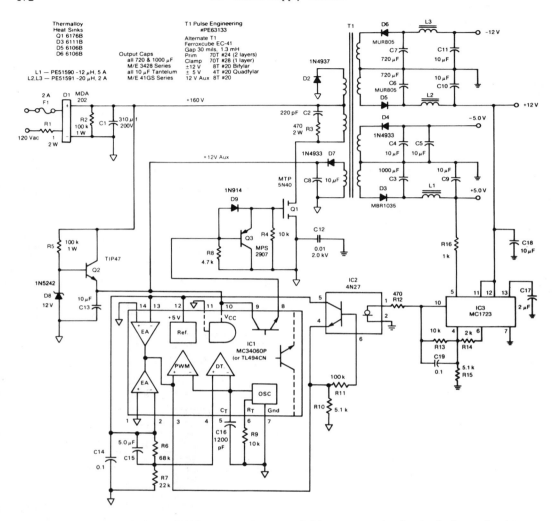

Two-Voltage, Dual Polarity, 60-W, Switching Power Supply. Provides 5- and 12-V DC outputs with both positive and negative polarities at 60 W from a 120-V AC input. Efficiency is 75% at a switching rate of 100 kHz. Output regulation is pro-vided by MBR1035 Schottky diodes at the 5-V output and MUR805 rectifiers at the 12-V output. Proper heatsinking is required where indicated.—"A 60 W, 100 kHz FET Switcher," TDT-101B, Motorola, Inc.

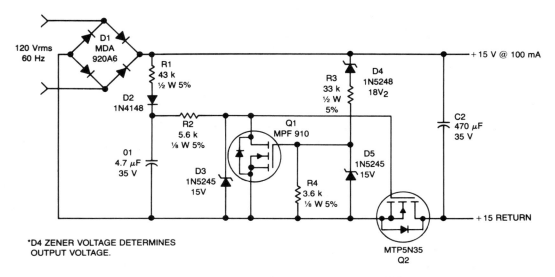

*D4 ZENER VOLTAGE DETERMINES
OUTPUT VOLTAGE.

Minimum Component Converter. Unregulated power source is non-isolated, but has minimum component count and is suitable for such applications as Nicad battery chargers, battery eliminators, industrial equipment, etc. Circuit operates by conducting only during the low voltage portion of the rectified sine wave. R1 and D2 charge C1 to approxi-

mately 20 V, which is maintained by Q1. This voltage is applied to the gate of Q2, turning it on. When the rectified output voltage exceeds the Zener voltage of D4, Q1 turns it on, shunting the gate of Q2 to ground, turning it off.—W. Skelton, "1.5 W Offline Converter," *TMOS Power FET Design Ideas*, p. 39, Motorola, Inc.

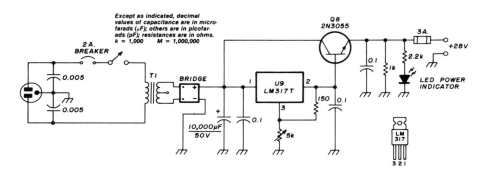

28-V Power Supply. Delivers 28 V DC at up to 2.5 A from a 120-V AC source. The output of transformer T1 is bridge rectified, filtered, and regulated by pass transistor 2N3055. Adjustable regulator U9 (LM317T) drives the base of the 2N3055 to set out-

put voltage and provide additional electronic filtering. T1 has a secondary providing 25.2 V at 2 A. The bridge rectifier is rated at 4 A and 100 PIV.—R. Littlefield, A Compact 75-Meter Monoband Transceiver, *ham radio*, November 1985, pp. 13–27.

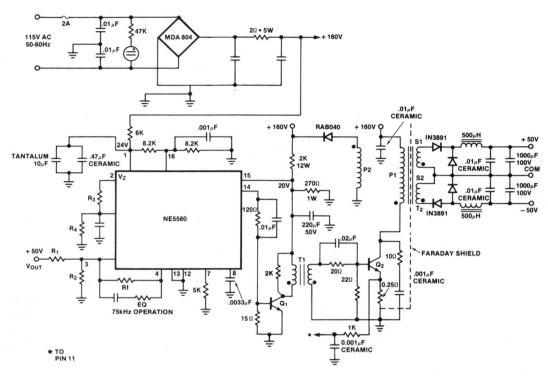

Dual Polarity 50-V Supply. NE5560 push-pull regulator device is operated at a switching frequency of 75 kHz to provide dual polarity 50-V outputs at 1 A. T1 has a primary of 60 turns of #24 wire on a Ferroxcube #2616 core with a secondary of 20 turns of #26 wire. T2 provides a 2.4:1 stepdown ratio; primary should be 60 turns of #26 wound between the two secondaries of 25 turns of #20 on a Ferroxcube 3622 core.—"Forward Converter Application Using the NE5560," AN121, Signetics Corporation.

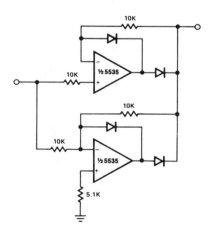

Precision Full Wave Rectifier. Provides accurate full wave rectification. Output impedance is low for both input polarities, and errors are small at all signal levels. Any load applied should be referenced to ground or a negative voltage. Reversal of all diode polarities will reverse the polarity of the output.— "Applications Using the SE/NE5535," AN150, Signetics Corporation.

19

Radioteleprinter Circuits

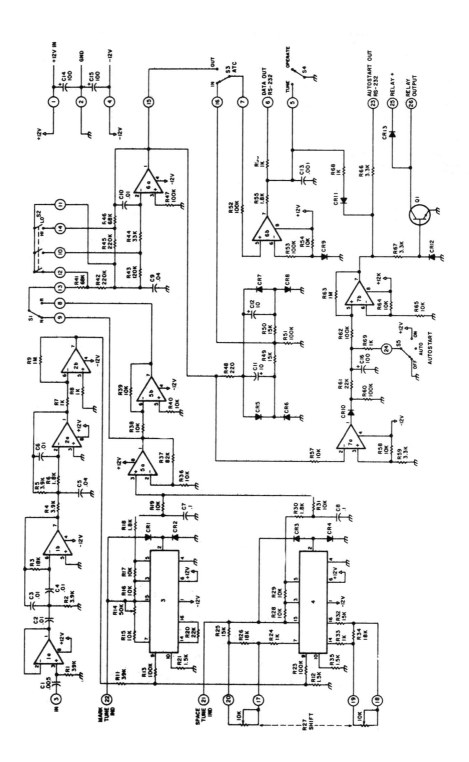

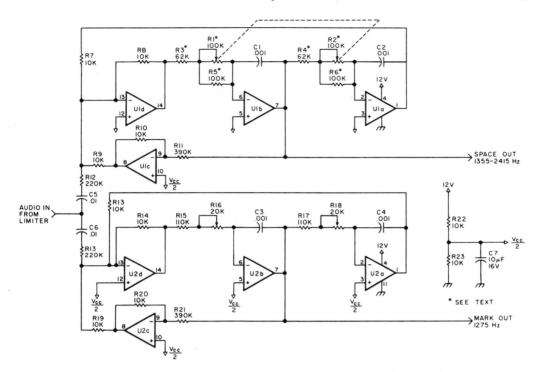

RTTY Filter.   Audio filter for RTTY reception is continuously tunable for shifts from 80–1160 Hz, with gain and Q nearly flat across this range. The "mark" filter is fixed at 1275 Hz while the "space" filter is tunable from 1355–2415 Hz. R3 and R4 should be closely matched, as should R5 and R6; these resistors determine the high and low ends of the tuning range.—D. Williams, "The Perfect RTTY Filter," 73 *Magazine*, January 1986, pp. 46–47.

◀ RTTY Terminal Unit.   Continuously adjustable shift from 50–850 Hz allows reception of RTTY signals. RS-232 data output interfaces with contemporary terminals, microprocessors, code converters, or current-loop drivers. Tuned circuits using active filter technology and input filtering ahead of the limiter reduces unwanted signals; a squelch circuit clamps the data output in the "mark" state when no data is present. IC1A is a combination low pass filter and follower, while IC1B and IC2A form high- and low-pass filters acting as broad bandpass filters. IC2B is a limiter stage with a gain of 1000. The channel filters are IC3 for "mark" and IC4 for "space." IC1, IC2, IC5, IC6, and IC7 are all sections of MC1458 dual op amps. IC3 and IC4 are AF-100-1CN. Q1 is a 2N3904 or equivalent switching transistor. CR1 to CR9 are 1N270 or equivalent germanium diodes. CR10 to CR12 are 1N914, while CR13 is a 1N4000. Original article includes details on construction, alignment, and use of the circuit.—B. Roehrig, "Tune In the TU-1000," 73 *Magazine*, June 1985, pp. 14–22.

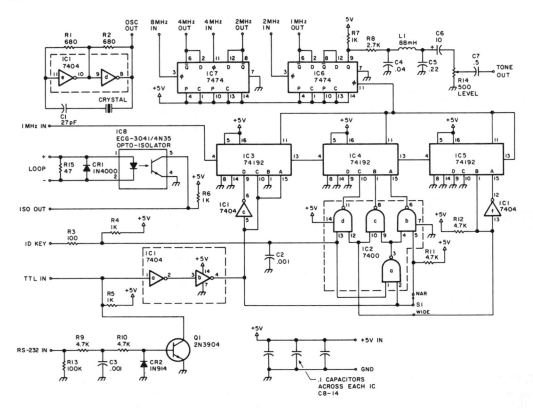

AFSK Oscillator. Crystal-controlled synthesized AFSK oscillator produces tones for RTTY. Circuit uses a standard source frequency of 1, 2, 4, or 8 MHz that can be derived from an on-board oscillator or taken from a microcomputer, frequency counter, or other standard frequency generator. Circuit will accept current-loop, TTL, or RS-232 inputs and produce a clean sine wave output.—R. Roehrig, "AFSK, And Ye Shall Transceive," 73 *Magazine,* June 1985, pp. 32–33.

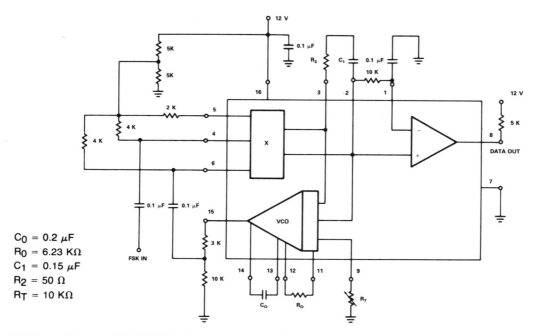

$C_0 = 0.2\ \mu F$

$R_0 = 6.23\ K\Omega$

$C_1 = 0.15\ \mu F$

$R_2 = 50\ \Omega$

$R_T = 10\ K\Omega$

FSK Demodulator. XR-210 PLL IC is used in FSK demodulator which recognizes 1070 Hz as the "mark" frequency and 1270 Hz as the "space" frequency. The free-running frequency of the PLL should be 1170 and potentiometer RT (10 K in this case) is adjusted to produce it. Operating parameters depend upon the values of C0, R0, C1, R2, and RT; original applications note includes design equations.—"XR-210/XR215/XR-S200 Phase Locked Loops," AN-22, Exar Corporation.

FSK Decoder. 300-baud FSK signals at 1070 and 1270 Hz are demodulated by the NE565 PLL IC. As a signal appears at the input, the loop locks to the input frequency and tracks it between the two frequencies with a corresponding DC shift at the output.—"Typical Applications With NE565," AN184, Signetics Corporation.

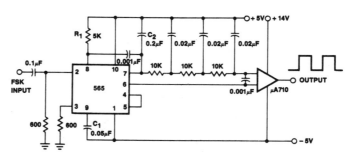

20

Receiving Circuits

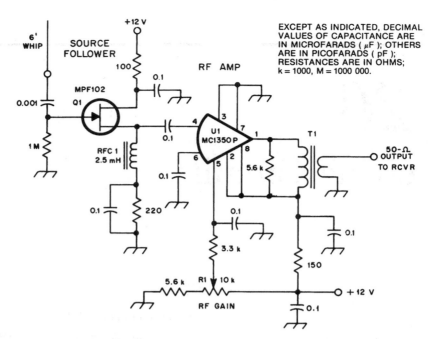

EXCEPT AS INDICATED, DECIMAL
VALUES OF CAPACITANCE ARE
IN MICROFARADS (μF); OTHERS
ARE IN PICOFARADS (pF);
RESISTANCES ARE IN OHMS;
k = 1000, M = 1000 000.

Active Antenna. A physically short antenna element (2 meters or less in length) followed by a high gain amplifier (40 dB or more of gain) enables performance equivalent to many outdoor "longwire" systems from 1.8–30 MHz. Intended for indoor use when an outside antenna is not permitted or practical, the antenna element delivers different signal-levels when placed in a horizontal or vertical position; the best position for a given frequency should be determined experimentally. Gain is controlled by potentiometer R1. RFC1 is a miniature 2.5-mH RF choke. T1 has a primary consisting of 30 turns of #28 wire on a FT50-43 ferrite toroid core and a secondary of four turns of #28 wire. If overloading from local AM broadcast band stations is experienced, a high pass filter with a cutoff frequency of 1.6 MHz should be added between the antenna element and the amplifier input.—D. DeMaw, "Learning to Work with Preamplifiers," QST, May 1986, pp. 21–23.

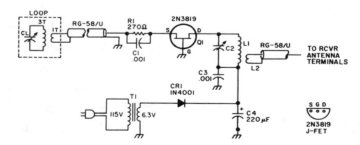

Loop-Antenna Amplifier. Designed for use with a loop receiving-antenna operating on the 160-meter (1800–2000 kHz) amateur radio band, but is sufficiently "broadband" in design to function on the AM broadcast band (540–1600 kHz) and the 80-meter amateur band (3500–4000 kHz) if the loop antenna is modified to cover those ranges. The 2N3819 FET is used in a grounded-gate configuration to deliver a gain of approximately 20 dB. The amplifier must be mounted in a shielded metal box and care must be taken to prevent "re-radiation" of the amplifier's output into the loop antenna itself.—J. Pepper, "The Hula Hoop Loop Revisited," 73 Magazine, May 1986, pp. 42–43.

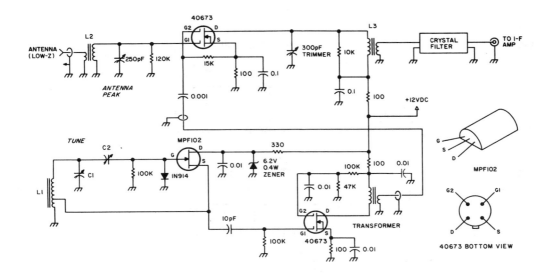

L4–50 TURNS #28 ENAMELLED WIRE ON AMIDON T50-2
5 TURN LINK TO DETECTOR

CRYSTAL–3393.6kHz FOR USB
3396.4kHz FOR LSB (USED HERE)

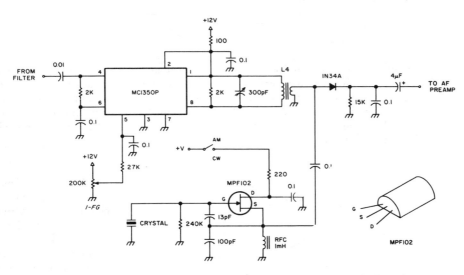

7.8–12.1 MHz Receiver. Covers 7.8–12.1 MHz and receives SSB and CW signals in addition to AM. Receiver is designed in separate mixer/filter/VFO and detector/IF amplifier/BFO sections. A single, tuned circuit matches the antenna to the mixer stage; a dual gate MOSFET is used as the mixer. The VFO uses a JFET in a Hartley oscillator and another dual gate MOSFET as a buffer amplifier. The output of the mixer is fed through a tuned circuit with a link output to a Heath filter of 2.1-kHz bandwidth and a 3.395-MHz center frequency. The IF amplifier is a single IC which provides more than 40 dB of gain.

The detector is a single diode and a BFO is provided by a Colpitts oscillator. Output of this circuit should be fed to an audio stage. C1 is a 170-pF variable capacitor. C2 is a 1.5–5.4-pF air-trimmer capacitor. L1 is 44 turns of #28 wire on a T50-6 form, tapped 12 turns above ground. L2 is 24 turns of #28 wire on a T68-2 form with a three-turn link to the antenna. L3 is 50 turns of #28 on a T50-2 form with a 22-turn link to the filter. The XFMR is 18 turns of #28 on a FT37-43 with a three-turn link to the mixer.—S. Mann, "The 30 Meter Plus Receiver," 73 *Magazine*, July 1985, pp. 30–32.

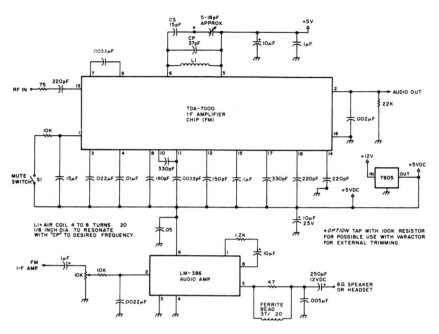

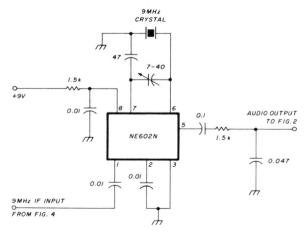

30–70 MHz IF Amplifier Strip. Suitable for use as the IF amplifier section of a microwave receiver or as a test bench IF amplifier. Heart of the unit is a TDA-7000 FM IF amplifier IC from Signetics. The local oscillator coil, L1, is wound from five turns of #26 wire on a ⅛-in. diameter form. The coil should be shielded and the shield grounded. Some adjustment in the values of the capacitors connected to L1 may be necessary for proper operation.—C. Houghton, "Microwave Building Blocks: The IF Amplifier," 73 *Magazine*, October 1986, pp. 42–44.

Product Detector. Signetics NE602 serves as product detector for 9-MHz IF signals. Power voltage for the NE602 is 6 V, supplied through a 1.5-K resistor to pin 8. Recommended 9-MHz IF signal level is 400 mV.—C. Klinert, "Build a Pocket Portable SSB Receiver," *ham radio*, November 1986, pp. 55–63.

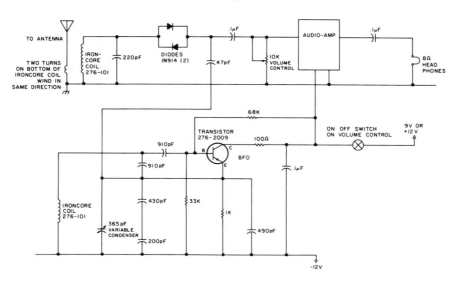

3500–4000 kHz One-Transistor Receiver. Circuit can be made to cover other frequency ranges in the shortwave spectrum by adjusting the capacitance and inductance values used. General purpose transistors such as the 2N2222 may be used in the place of de- vice specified in diagram. The inductance values of the coils are approximately 10 mH.—K. Hand, "One Band, One Transistor," 73 *Magazine*, October 1986, p. 54.

10-kHz to 30-MHz Active Antenna. Amplifier compensates for short antenna-element and delivers 20-mW output into a 50-ohm output. IMD performance characteristics are better than most conventional active antenna designs. Antenna element must be placed outside of metal structures.—R. Burhans, "Active Antenna Preamplifiers," *ham radio*, May 1986, pp. 47–54.

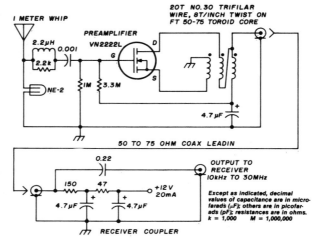

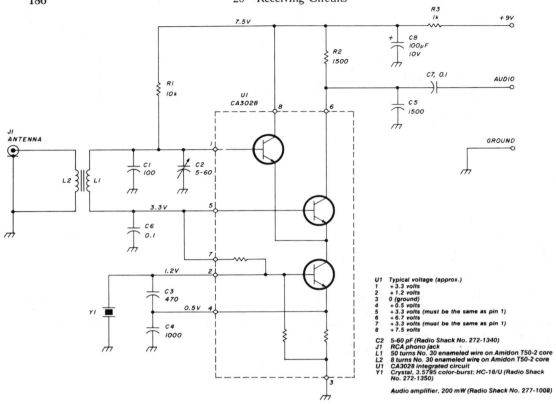

U1   Typical voltage (approx.)
1   + 3.3 volts
2   + 1.2 volts
3   0 (ground)
4   + 0.5 volts
5   + 3.3 volts (must be the same as pin 1)
6   + 6.7 volts
7   + 3.3 volts (must be the same as pin 1)
8   + 7.5 volts

C2   5–60 pF (Radio Shack No. 272-1340)
J1   RCA phono jack
L1   50 turns No. 30 enameled wire on Amidon T50-2 core
L2   8 turns No. 30 enameled wire on Amidon T50-2 core
U1   CA3028 integrated circuit
Y1   Crystal, 3.5795 color-burst; HC-18/U (Radio Shack
     No. 272-1350)

Audio amplifier, 200 mW (Radio Shack No. 277-1008)

Simple 3580-kHz Receiver.   Designed to receive code practice and bulletins transmitted in CW on 3580 kHz from amateur radio station W1AW. As designed, circuit used 3.5795-kHz "color burst" crystal; other frequencies in the 3500–4000 kHz can be received by using different crystals. Heart of the receiver is a CA3028 IC; internal diagram of this device is shown in schematic. L1 and L2 are wound on the same T50-2 toroid core. L1 is 50 turns #30 wire while L2 is eight turns of #30.—E. Gellender, "A Simple 80-Meter Receiver," *ham radio*, November 1986, pp. 25–29.

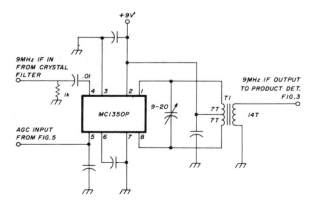

9-MHz IF Amplifier.   MC1350 IC is heart of IF amplifier which features high gain, excellent stability, and an AGC output terminal. Gain is 50 dB with a 200-μV IF output for a 0.7-μV input at 9 MHz. T1 has a center-tapped primary of #27 wire on a FT37-61 form with a 14-turn secondary of #27. —C. Klinert, "Build a Pocket Portable SSB Receiver," *ham radio*, November 1986, pp. 55–63.

Automatic Gain Control. The 2N2907 amplifies earphone audio which is half-wave rectified by the 1N4001. The time constant circuit uses a 10-μF capacitor and a 500-K resistor. The 2N2222 is a DC amplifier that drives the MC1350. A 100-μF electrolytic capacitor was used to bypass the 9-V power line to prevent low frequency oscillations.—C. Klinert, "Build a Pocket Portable SSB Receiver," *ham radio*, November 1986, pp. 55–63.

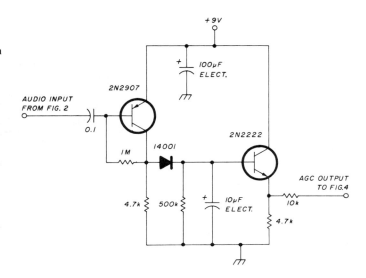

5-MHz First Mixer. L1 tunes mixer stage from 5000–5375 kHz. Heart of mixer is NE602; two LC circuits are also used. L1 is a 0.375-in. iron slug-tuned coil wound with 17 turns of #27 wire. L2 is a small RF choke. L3 is a 0.250-in. iron slug-tuned form with 14 turns of #27 wire. The input coupling link is five turns of insulated "hook-up" wire. —C. Klinert, "Build a Pocket Portable SSB Receiver," *ham radio*, November 1986, pp. 55–63.

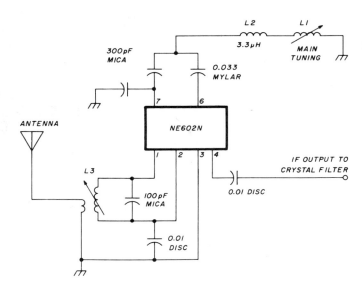

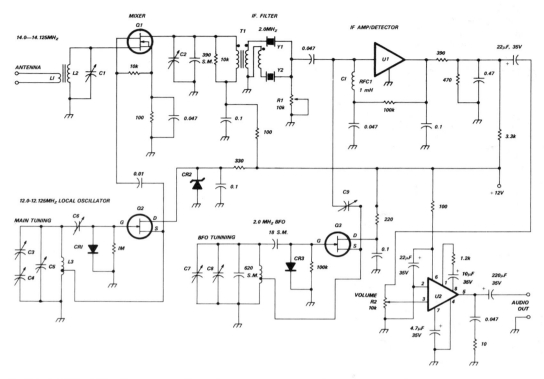

14,000–14,350 kHz Superheterodyne Receiver.
Use of a ZN414 IF amplifier/detector IC provides
sharp selectivity and a vast improvement over direct-
conversion-receivers. After passing through a fixed-
tuned preselector, incoming signals at 14 MHz are
admitted directly to gate 1 of the dual gate MOSFET
mixer. Mixer output at the IF is passed through a
tuned circuit at the primary of T1. The output of T1
is supplied to the crystal filter as two signals of opposite phase. The effect is that the parallel capacitances
within the two crystals are cancelled, leaving an extremely high-Q series-resonant circuit (a "half-
lattice" filter). The local oscillator and BFO are identical except for the tuned circuits. C1, C2, C5, and
C8 are 15–150 pF trimmer capacitors while C3 is a
15–90 pF air variable capacitor with vernier drive.

C4 is a 5–60 pF trimmer. C6 and C9 are 3–10 pF
trimmers. C7 is a 1.4–13 pF air variable capacitor.
CR1 and CR2 are 1N914 diodes while CR3 is a
6.2-V Zener diode. L2 is 22 turns of #24 on a T50-6
form; L1 is two turns around the ground end of L2.
L3 is 24 turns of #24 on a T-50-6 tapped six turns
from ground end. L4 is 40 turns of #24 on a T-68-2
form tapped 10 turns from ground end. Q1 is a
40673 dual gate MOSFET and Q2/Q3 are MPF102
FETs. R1 is a 10-K "trimpot" and R2 is a 10-K potentiometer. T1 is 12 bifilar turns of #24 wire on a
FT-37-61. U1 is a Ferranti ZN414 and U2 is a
LM386. Y1 and Y2 are 2-MHz crystals in HC6/U
holders.—E. Bodner, "20-Meter HF Superhet,"
ham radio, November 1986, pp. 65–72.

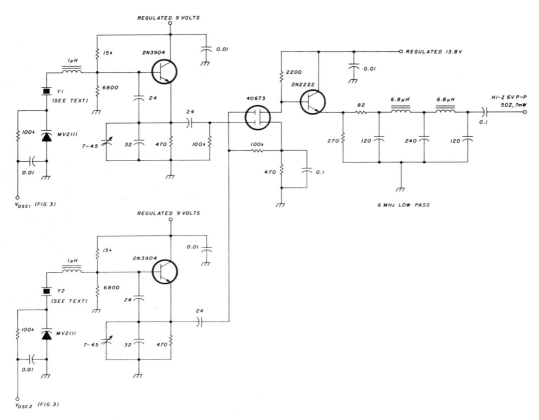

High Stability BFO Using Dual VXOs.   Twin Colpitts oscillators use varactor tuning to provide a variable-beat-frequency source from 5593–5597 kHz. Crystals Y1 and Y2 can be in the 10–22 MHz range, operated in their fundamental mode, although 15-MHz crystals were used in circuit shown. It is important that the two crystals be as closely matched in frequency as possible. The BFO shift can be extended from 4–6 kHz by installing 1-μH inductors in series with the crystals. The two oscillators are followed by a mixer stage and buffer.—P. Bertini, "A High Stability BFO for Receiver Applications," *ham radio*, June 1985, pp. 28–31.

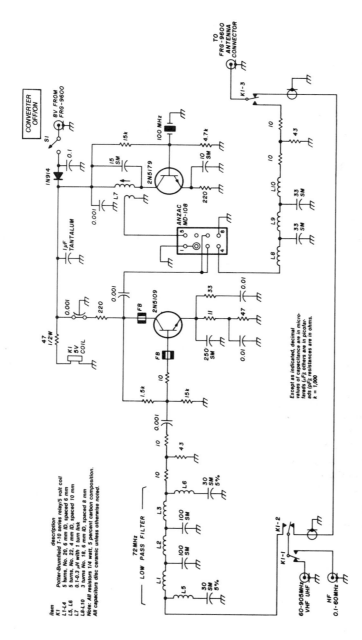

0.1–60 MHz to 100.1–160 MHz Receiving Converter. Designed to allow VHF receivers to tune lower frequency ranges. Anzac MD-108 double-balanced mixer module is used to reduce unwanted mixing products. Circuit was designed for use with Yaesu FRG-9600 receiver, and power for the circuit is obtained from an auxiliary jack on the FRG-9600, which can deliver 8 V at 200 mA, more than adequate to power this circuit. K1 is a Potter-Brumfield T-10 series relay. L1–L4 are all five turns of #20 wire, 6 mm in diameter, spaced 6 mm. L5 and L6 are five turns of #22 wire, 4 mm in diameter, spaced 10 mm. L7 is 0.1–0.3 μH with a one-turn link. L8 to L10 are three turns of #18, 6 mm in diameter, spaced 8 mm.—E. Guerri, "Add General Coverage to Yaesu's Latest VHF-UHF Receiver," *ham radio*, October 1985, pp. 67–70.

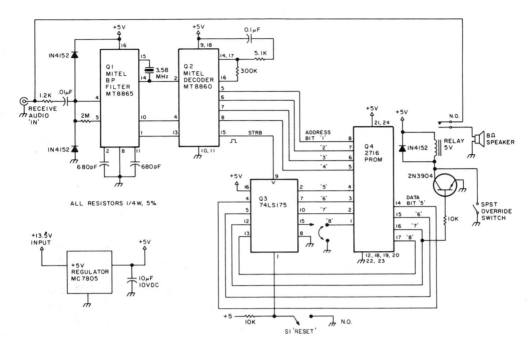

Selective Calling. Variation of a DTMF decoder silences a receiver until the proper sequence of touch-dialing tones is received. Heart of system is a Mitel MT8860 DTMF decoder IC and its matching MT8865 bandpass filter; operation of the MT8860 is controlled by the program stored in the 2716 PROM.

Original article includes a program listing for the 2716 and details on programming the desired tone-sequence for accessing the receiver.—N. Robin, "Secal and State Machines," 73 *Magazine,* January 1986, pp. 28–34.

1.8–29.7 MHz Antenna System and Tuner. Antenna portion consists of approximately 200 feet of wire, with 50 feet in a vertical plane and the remainder running horizontally to a nearby tiepoint; the antenna works against ground in Marconi fashion. L1 is a roller inductor, 35 turns and 2.5-in. in diameter. L2 and L3 are B&W 3900, 32 turns each, 2-in. in diameter, eight turns per in., tapped every six turns. Tuner is adjusted for best performance of tuner element.—B. Orr, "Ham Radio Techniques," *ham radio,* August 1986, pp. 43–46.

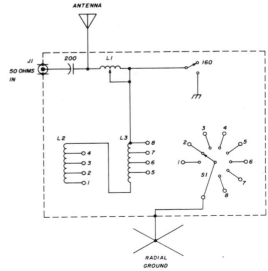

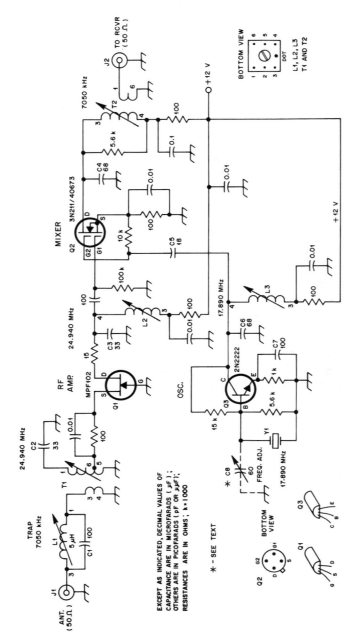

24-MHz to 7-MHz Receiving Converter. Converts signals received on the new 24,890–24,990 kHz amateur radio band to signals on 7000–7100 kHz (also an amateur band) for reception on receivers not covering the recently added band. Circuit uses a MPF102 JFET in a grounded-gate configuration for stability. A LC "trap" with a center frequency of 7050 kHz is used at the antenna input to prevent interference from other signals in the 7000–7100 kHz range. T1's secondary is nine turns of #26 wire wound on a L57-6 form, tapped two turns above ground. The primary is a one-turn winding over the secondary. T2 is 22 turns of #26 on a L57-6 form with a secondary of three turns. L1 is 26 turns of #30 on a L43-6 form, L2 is 10 turns of #26 on a L43-6 form, and L3 is 12 turns of #26 on a L43-6 form.—D. DeMaw, "A Converter for the 24 MHz WARC Band," *QST*, April 1985, pp. 42–44.

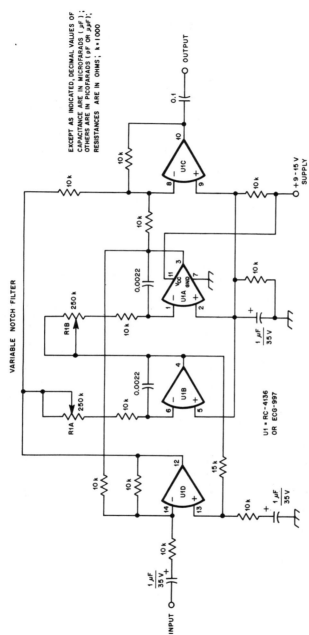

Variable Notch Filter

EXCEPT AS INDICATED, DECIMAL VALUES OF CAPACITANCE ARE IN MICROFARADS (µF) ; OTHERS ARE IN PICOFARADS (pF or µµF); RESISTANCES ARE IN OHMS ; k = 1000

U1 = RC-4136 OR ECG-997

Variable Audio Notch Filter. Provides a tunable rejection "notch" from 300–2800 Hz in the audio output of a receiver for removal of interfering heterodynes, CW signals, etc. R1 is a dual 250-K linear-taper potentiometer while U1 is a quad op amp IC such as the RC4136, ECG997, or equivalent. When not in use, tune R1 so that the rejection notch is outside the audio range of the receiver's output. Circuit is generally most effective during reception of CW signals.—T. Desauliniers, "Variable Notch Filter for Receivers," *QST*, January 1985, pp. 39–40.

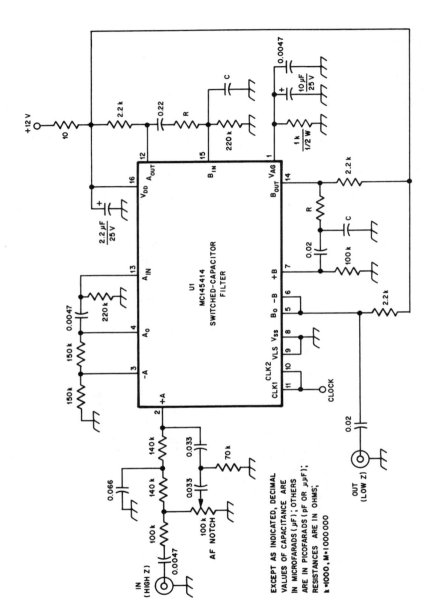

Switched-Capacitor Audio Filter. Based on the MC145414 switched-capacitor filter IC, this circuit offers a tunable notch and cascaded low pass filters with output buffering. Notch filter covers 40–4000 Hz. The low pass filter's cutoff frequencies are determined by the frequency of the clock signals; a 100-kHz clock gives a cutoff of 2800 Hz, 50-kHz gives 1400 Hz, and 25-kHz gives 700 Hz. Clock signals to pins 10 and 11 must be square waves. The values of R and C should be determined experimentally to remove any remaining clock signal "residue."—R. Schellenbach, "A New Switched Capacitor Filter IC from Motorola," *QST*, January 1985, pp. 42–43.

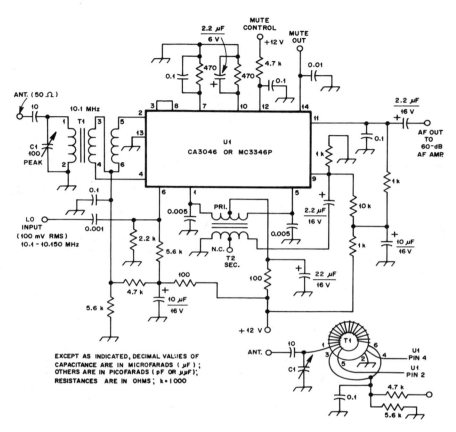

Direct-Conversion-Receiver Front End. Circuit uses the CA3046 transistor array IC rather than discrete transistors. The configuration provides a singly balanced product detector which minimizes unwanted AM detection. The conversion gain is on the order of 10 dB; a RF amplifier is not needed ahead of the detector on frequencies below 14 MHz. The mute control-terminals are provided if the receiver is used in conjunction with a transmitter. T1 has a primary of 24 turns of #26 wire wound on a T50-2 toroid core with a secondary of 10 bifilar turns of #26 wire spread over the primary winding. T2 is a miniature audio transformer with a 10-K primary and 2-K secondary; secondary center tap is not used.—D. DeMaw, "A Utility IC—The CA3046," *QST*, August 1985, pp. 21–24.

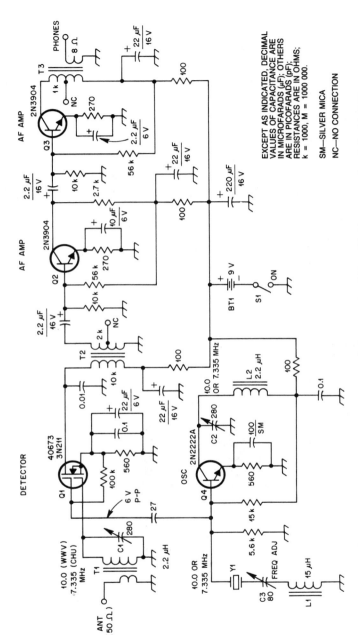

Fixed-Frequency HF Receiver. Originally designed for reception of standard time and frequency stations WWV and CHU, circuit can also be used for reception of any single frequency between approximately 5–15 MHz. Frequency is determined by Y1, a fundamental-frequency crystal with an 30-pF load capacitance. Circuit is also capable of satisfactory SSB and CW reception. Q1 is the detector while Q4 is the BFO for SSB/CW reception. A LM386 audio amplifier IC can be added if output from a speaker is desired; if this is done, a 10-K potentiometer should be added between Q3 and the LM386 as a volume control. L1 is 16 turns of #26 wire on a FT37-61 ferrite toroid. T1 has a primary of two turns of #26 wire wound on a T68-6 toroid and a secondary of 22 turns of #26. T2 is a miniature audio transformer with a 10-K primary and 2-K secondary. T3 is an audio output transformer with a 1-K primary and an 8-ohm secondary.—D. DeMaw, "WWV and CHU in Your Workshop," *QST*, October 1986, pp. 42–44.

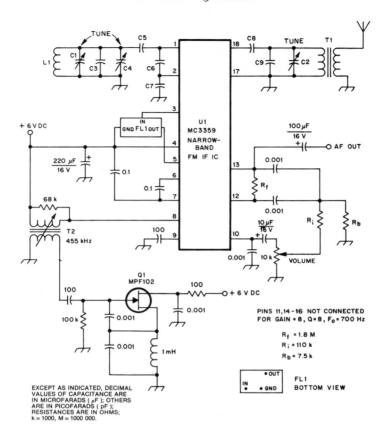

EXCEPT AS INDICATED, DECIMAL
VALUES OF CAPACITANCE ARE
IN MICROFARADS (μF); OTHERS
ARE IN PICOFARADS (pF);
RESISTANCES ARE IN OHMS;
k = 1000, M = 1000 000.

PINS 11,14 - 16 NOT CONNECTED
FOR GAIN = 8, Q = 8, F_0 = 700 Hz

R_f = 1.8 M
R_i = 110 k
R_b = 7.5 k

FL1
BOTTOM VIEW

Simple Receiver for 7 or 10.1 MHz. Motorola MC3359 narrow band FM IF IC is adapted for use in a simple receiver covering the 7 or 10.1 MHz amateur radio bands; accompanying table shows proper parts values for desired frequency coverage. FL1 is a ceramic bandpass filter whose bandwidth depends upon the signal mode desired for reception. If SSB or CW reception is desired, a BFO signal will need to be supplied at pin 8 of the MC3359. An ECG860 IC may also be substituted for the MC3359. Audio output is sufficient to drive an earphone or sensitive headphones.—B. Williams, "The SIMPLEceiver," *QST*, September 1986, pp. 34–39.

Circuit Elements for Both 30 and 40 Meters

30 Meters

C1, C2	C3	C4	C5	C6	C7	C8	C9	L1	T1
3-10	100	0-5	120	180	180	15	100	13t no. 26 enam on T25-6 toroid	*Pri*—16t no. 26 enam on T50-6 toroid. *Sec*—4t no. 26 enam on primary

40 Meters

C1, C2	C3	C4	C5	C6	C7	C8	C9	L1	T1
5-60	180	0-100	120	390	390	25	180	16 t no. 26 enam on T37-6	*Pri*—36t no. 26 enam on T50-6. *Sec*—4t no. 26 enam over primary

Note: All capacitor values are in picofarads.

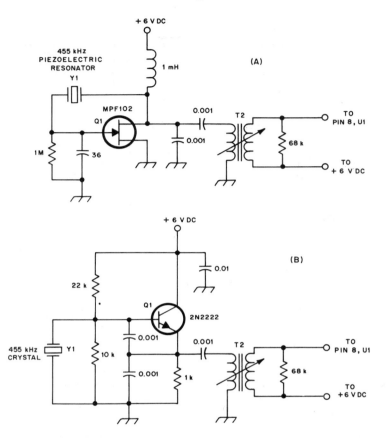

Beat Frequency Oscillators. Two modified Pierce-oscillator circuits reflect different approaches to providing a beat frequency signal at 455 kHz for reception of SSB and CW signals. The circuit at A uses a 455-kHz piezoelectric resonator in place of a crystal. However, it may be difficult to get the circuit to os-cillate properly. The circuit at B is a more conventional LC oscillator using a 455-kHz crystal. T2 in both circuits is a 455-kHz miniature IF transformer.—B. Williams, "The SIMPLEceiver," *QST*, September 1986, pp. 34–38.

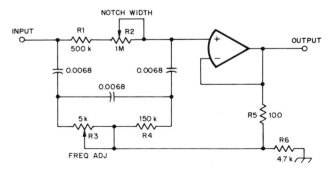

Notch Filter for Direct Conversion Receivers. Active bridge-differentiator circuit provides a notch tunable from 400–2000 Hz at a depth of 40 dB. Op amp can be most general purpose devices such as the LF353.—P. Kranz, "Better Ears for the MAVTI-40 Transceivers," *QST*, October 1985, pp. 14–20.

WWV Converter.   Converts 10-MHz signal of standard time and frequency station WWV to 1.5 MHz for reception on ordinary AM broadcast band radios. The MOSFETs are used in a dual gate configuration as a mixer and is triode-connected in a crystal oscillator circuit.— "Using MOSFET Integrated Circuits in Linear Circuit Applications," AN-4590, GE/RCA Solid State.

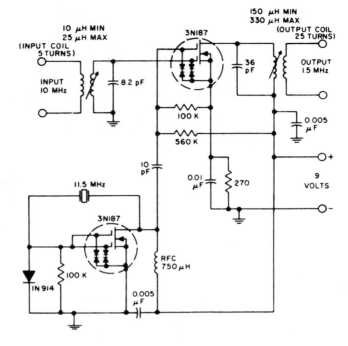

FM Demodulator.   Uses 4046 PLL IC to demodulate a 10-kHz carrier frequency modulated by a 400-Hz audio signal. Total FM signal amplitude is 500 mV. The center frequency is determined by the values of C1 and R1.—"The RCA COS/MOS Phase Locked Loop: A Versatile Building Block for Micropower Digital and Analog Applications," ICAN-6101, GE/RCA Solid State.

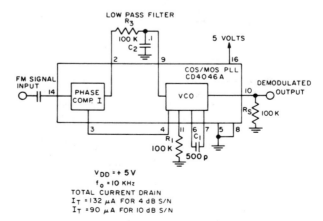

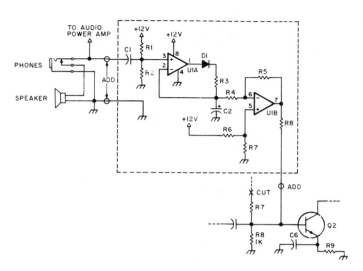

Automatic Gain Control.   Helps hold audio output of a receiver at a relatively constant level by sampling the signal present in the audio stages of a receiver and adjusting the gain of the IF stages in response. In circuit shown, audio sample is taken from the receiver's headphone jack. The circuit detects the audio level and uses it to reduce the voltage of the final transistor in the receiver's IF stage (Q2 in the illustration) if the audio level suddenly increases; the reduced voltage of Q2 reduces the gain of the IF amplifier. Capacitor C1 is a 0.01-μF ceramic while C2 is a 10-μF tantalum. R1, R2, R6, and R7 are all 47 K. R3 is 1 K, R4 is 100 K, R5 is 220 K, and R8 is 2.7 K. D1 is either a 1N914 or 1N4148, while U1 can be any common FET input op amp such as the LF412 or TL084. —S. McArthur, "AGC, PDQ," 73 *Magazine*, May 1986, pp. 66–67.

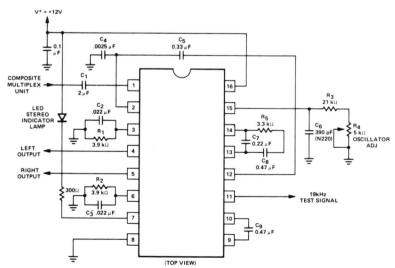

FM Stereo Multiplex Decoder.   Accepts a composite FM stereo multiplex signal and delivers separate left and right channel outputs. Heart of the circuit is a 758 PLL device. LED lights when a stereo multiplex signal is input.—"FM Stereo Multiplex Decoder," 1985 Linear LSI Data and Applications Manual, 5-80, Signetics Corporation.

SCA Demodulator. Recovers the SCA (subsidiary carrier authorization) signal carried as a 67-kHz FM subcarrier, above the frequency range of normal programming, by many commercial FM stations. SCA programming is typically background music or special interest material such as "talking books" for the blind. NE565 PLL IC is used; circuit should be connected between the FM discriminator and de-emphasis filter of existing FM receiver.—"Typical Applications With NE565," AN184, Signetics Corporation.

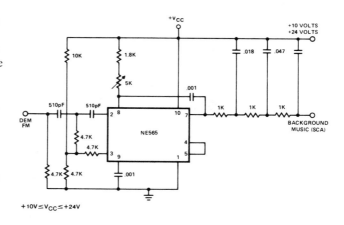

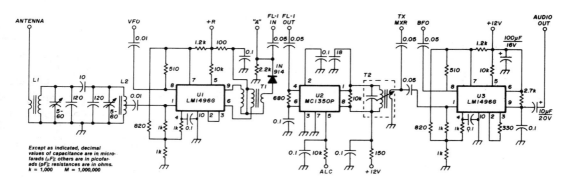

3800–4000 kHz Receiving Strip. Designed as receiver section of a SSB transceiver covering 3800–4000 kHz. IF frequency is 9 MHz and a 5-MHz VFO is used in the transceiver. IF stage U2 (MC1350P) is shared with the transmitter section; it offers 45-dB gain and has an AGC range of about 70 dB. U3 (LM1496G) is a product detector. L1 is 38 turns of #28 wire on a T50-2 form with a three-turn link. L2 is 38 turns of #28 wire on a T50-2 with a tap at 18 turns. T1 is nine turns of #28 × 3 on a FT37-43 form. T2 is a 10.7-MHz IF transformer.—R. Littlefield, A Compact 75-Meter Monoband Transceiver, *ham radio*, November 1985, pp. 13–27.

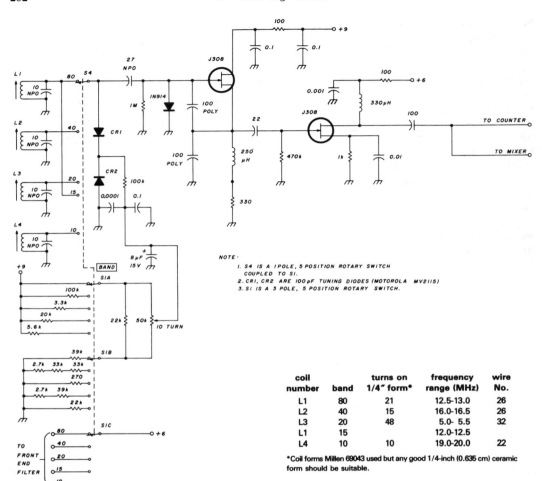

NOTE:
1. S4 IS A 1 POLE, 5 POSITION ROTARY SWITCH COUPLED TO S1.
2. CR1, CR2 ARE 100 pF TUNING DIODES (MOTOROLA MV2115)
3. S1 IS A 3 POLE, 5 POSITION ROTARY SWITCH.

coil number	band	turns on 1/4" form*	frequency range (MHz)	wire No.
L1	80	21	12.5-13.0	26
L2	40	15	16.0-16.5	26
L3	20	48	5.0- 5.5	32
L1	15		12.0-12.5	
L4	10	10	19.0-20.0	22

*Coil forms Millen 69043 used but any good 1/4-inch (0.635 cm) ceramic form should be suitable.

VFO for Receivers.   Varactor diode tuning is used to eliminate hard-to-find variable capacitors as well as the mechanical linkage they require. The varactors require DC of high purity. Rotary bandswitch selects coil for desired output; accompanying table gives coil data for various frequency ranges. Original article includes full information on circuit construction and alignment.—R. Thompson, 10 Through 80-Meter Homebrew Receiver, *ham radio*, November 1985, pp. 40–52.

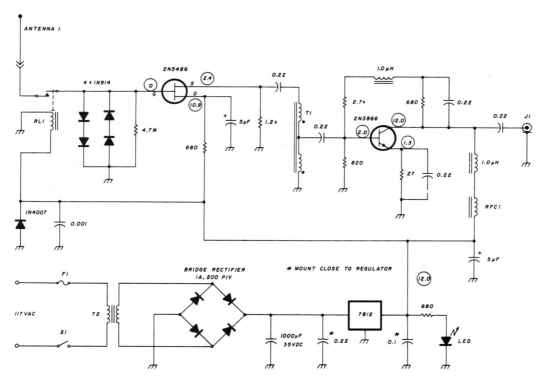

0.5–30 MHz Active Antenna. JFET transistors in RF amplifier section provide reception equal to most outdoor random wires from 0.5–30 MHz. Four 1N914 diodes help prevent damage to the 2N5486 from static electricity accumulation on the antenna element. A MPF102 may be substituted for the 2N5486. The reed relay (RL1) disconnects the antenna element from the preamp stages when the antenna is not in use to prevent TVI caused by diode rectification of RFI from nearby HF or VHF transmitters.—P. Bertini, Active Antenna Covers 0.5–30 MHz, *ham radio*, May 1985, pp. 37–43.

item	description
ANT-1	antenna probe, 20-50 inch length. Stainless steel antenna whip, aluminum tubing, or collapsible whip. Radio Shack No. 15-232 (38 inches) or No. 270-1401, 270-1402, 270-1403, or 270-1405.
F1	1 amp, fast-blow fuse
J1	RF connector, SO-239
RFC-1	20 turns No. 26 wire on Amidon FT50-61 core material (see text)
RL1	reed relay, SPST. 12 VDC coil.
S1	SPST switch, 125 VAC rating. Radio Shack 275-612 or equivalent
T1	9 bifilar turns, No. 26 wire on Amidon T68-8 core material. Wire 4-6 twists per inch (see text)
T2	power transformer, 117 VAC to 12.6 VAC at 100 mA minimum. Radio Shack No. 273-1385 or equivalent

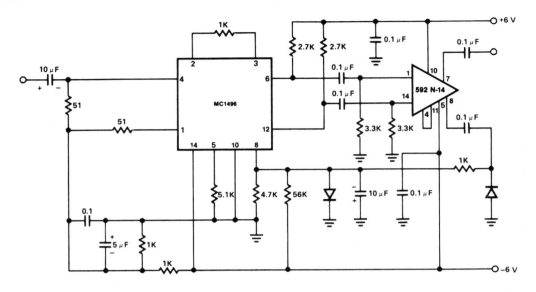

All resistor values are in ohms.

High Frequency Automatic Gain Control. NE592 video amplifier is used in conjunction with MC1496 balanced modulator to form AGC section. The NE592 has a bandwidth of up to 120 MHz and gains of up to 400. The input signal is fed to the MC1496 and coupled to the NE592 by a RC network. Rectify-ing and filtering one of the NE592 outputs produces a DC signal proportional to the AC signal amplitude. The signal is applied to the MC1496 as a control voltage.—"Using the NE592/5592 Video Amplifier," AN141, Signetics Corporation.

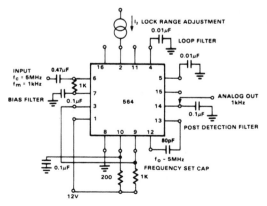

FM Demodulator for 12-V Operation. 564 PLL IC is used for FM detector. The input signal is AC coupled with the output signal being extracted at pin 14. Loop filtering is provided by the capacitor at pins 4 and 5. For best results, the frequency deviation in the input signal should be 1% or higher.— "SE/NE564 Phase Locked Loop," data sheet, Signetics Corporation.

21
Repeater-Station Circuits

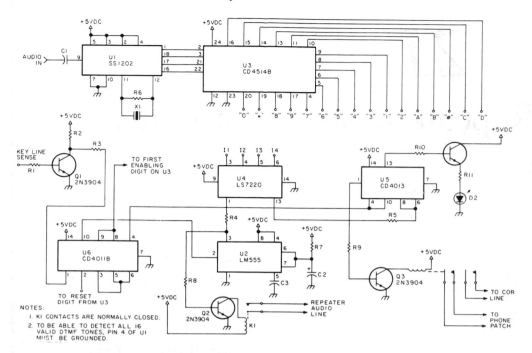

Repeater Autopatch. Connects amateur radio FM repeater station with telephone system. U1, a SSI202 from Silicon Systems, is a complete DTMF decoder with band-splitting filters. U4, a LS7220 by LSI Computer Systems, is a "keyless lock" IC used to ensure the security of the system by preventing operation of the circuit until the proper sequence of access tones is received. Output of circuit goes to any conventional "telephone patch" used to connect amateur radio stations to the telephone network.—P. Putnam, "The Piggy-Bank Patch," 73 *Magazine,* July 1986, pp. 40–43.

Subaudible-Tone Counter for Repeater Systems. Designed for use with a receiver to display the frequency of a subaudible tone used to access a repeater system. Frequency range covered is 67–250 Hz with an FM deviation of 0.5 kHz or less. The design of the bandpass filter and PLL is such that the loop "locks" in about 1 second if the subaudible tone has 0.5-kHz deviation. ICs U1-U4 make up the frequency display and are all TIL311. U5 is a 7805 and U6 is a 7808. U7-U10 are all 74C160. U11 is a 74LS74. U12 is a CD4011. U13 is a CD4060. U14 and U15 are both CD4518. U16 is a CD4046. U17 is a TL084. U18 is a MF10CN. The 4–34-pF ceramic trimmer capacitor is used to set the time base of the circuit. This can be measured at pin 13 of U14 and should be 0.2 seconds. Transmitting stations with poorly filtered power supplies may produce a reading of 120 Hz on the display.—P. Bunnell, "Subaudible Snooping," 73 *Magazine,* November 1986, pp. 48–49.

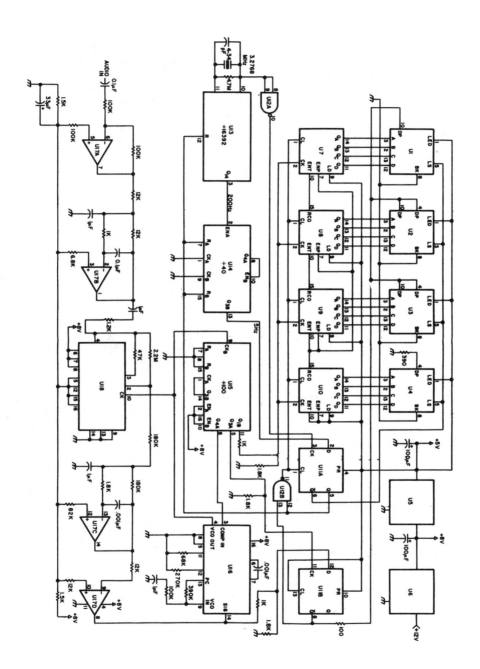

207

DTMF Controller for Repeater Systems. Allows control of an amateur radio repeater station through appropriate DTMF tones. Circuit features full DTMF decoding, multiple digit coding, latching output, multi-source reset, and a separate auxiliary output for use by system logic. Circuit has built-in separate DTMF decoder and sequence-detector sections located on separate circuit boards. The heart of the DTMF decoder is a SSI-201 DTMF decoder IC. A low pass filter consisting of R1, R2, C1, and C2 has been used. U2 is a 4-to-16 line decoder while U3 to U5 are inverting on non-inverting hex buffers. If a logic-1 output is desired, U2 should be a 14514 while a 14515 should be used if a logic-0 output is desired; if a 14515 is used, the hex buffers can be eliminated. The sequence detector consists of a register and a latch. Original article includes full details on construction and tailoring of the circuit to specific applications.—T. Simonds, "A DTMF Controller for Repeaters," *ham radio*, September 1985, pp. 47–54.

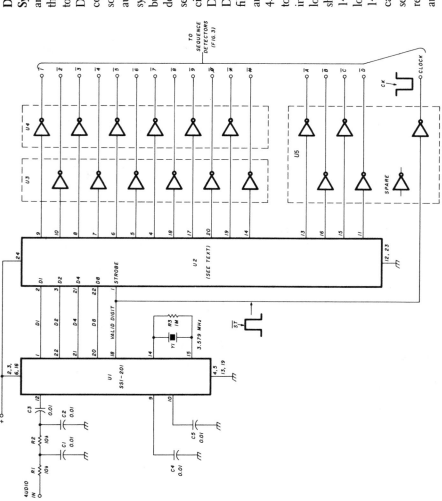

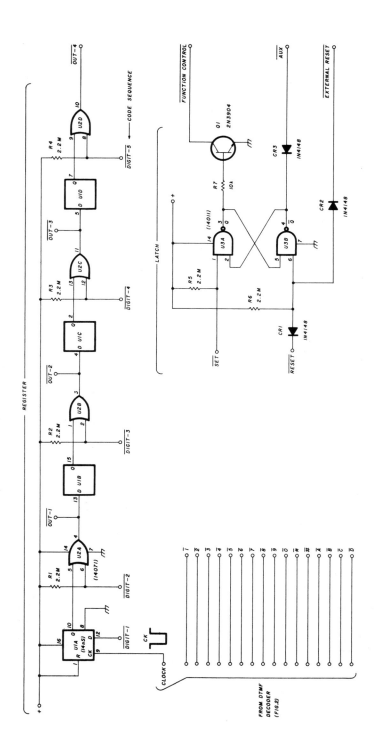

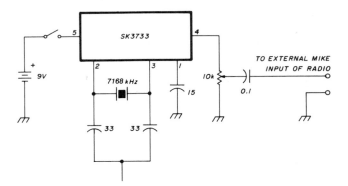

Tone-Burst Generator for Repeater Activation. Output of 7168-kHz crystal is divided down to produce a 1750-kHz audio output to activate amateur radio repeater stations requiring this audio frequency for access (particularly common in Europe). An ECG1197 may be substituted for the SK3733.—L. Nagurney, "A Tone Burst Generator for European Repeaters," *ham radio*, July 1986, p. 88.

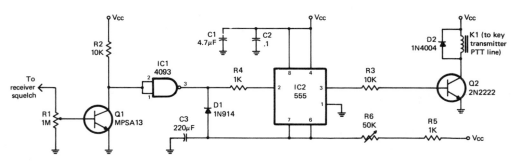

Parts List
COR

R1: 1 megohm PCB mount pot
R2, 3: 10K, ¼ watt
R4, 5: 1K, ¼ watt
R6: 50K PCB mount pot
Q1: 2N2222 NPN
Q2: MPSA-13 NPN-D
U : 4093 Schmitt Trigger
U2: 555 timer
C1: 4.7 uF electrolytic

C2: .1 disc cap.
C3: 220 uF electrolytic
D1: 1N914 diode
D2: 1N4004 diode
RY1: 12 VDC relay

Miscellaneous
PC board
Enclosure box
RG-174 cable

Carrier Operated Relay. Switches-on the transmitter section of an amateur radio repeater system when a signal is received on the system input-frequency. Input from the receiver's squelch circuit is input through a 1-M potentiometer; level should be set so that signal sufficient to "open" the receiver squelch is also adequate to activate the circuit.—J. Meyer, "How to Build a 2 and 10 Meter Repeater System On a Shoestring Budget," *CQ*, June 1985, pp. 48–50.

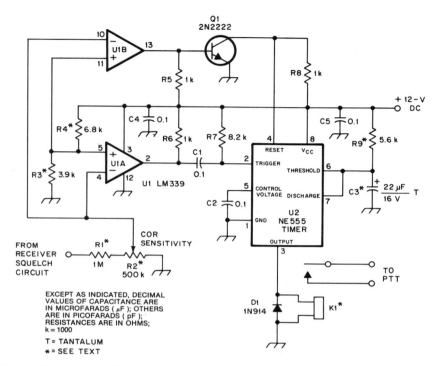

Carrier Operated Relay. Used in repeater circuits to switch on transmitter section when a signal is received on the system input frequency. A very small portion of the voltage from the receiver squelch circuit is sampled across R1 and R2. This sample is fed to pins 4 and 10 of U1, a LM339 quad comparator IC. The values of R3 and R4 may have to be adjusted for best results with a given repeater system. Diode D1 across relay K1 protects the 555 timer. Circuit as shown gives a timer cycle of slightly over 2 minutes. A different timing cycle is found by the formula (R9 × C3 × 1.1).—J. Arnold, "A Poor Ham's COR and Timer Circuit," *QST*, April 1986, p. 40.

22

Single Sideband Circuits

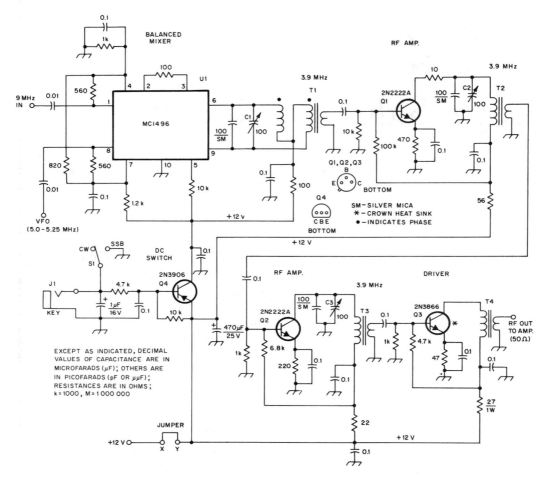

3.9-MHz Mixer and Amplifier Stage for SSB Transmitters. Accepts 9-MHz input signal from a local oscillator and a 5–5.25-MHz VFO signal to produce a 3.9-MHz SSB signal suitable for amplification. The MC1496 operates as a double-balanced mixer. Transformer T1 matches impedance between the MC1496 and transistor Q1 (2N2222A), which operates as a class A RF amplifier. Q2, another 2N2222A, also operates in a linear fashion and further amplifies the signal. Q3 (2N3866) also operates class A but the output contains a broadband transformer, T4. C1, C2, and C3 need to be "stagger tuned" (peaked at the high, center, and low frequencies of desired operating range). T1 is bifilar wound from #26 wire, eight twists per in., with a primary of 13 turns on a FT-50-61 ferrite toroid and a secondary of three turns over the primary winding. T2 has a primary of 13 turns of #26 on a FT-50-61 form and a secondary of nine turns. T3 has a primary of 13 turns of #26 on a FT-50-61 form and a secondary of four turns. T4 is 15 turns of #26 wire on a FT-50-43 ferrite toroid with a secondary of seven turns.—D. DeMaw, "The Principles and Building of SSB Gear," *QST*, November 1985, pp. 16–19.

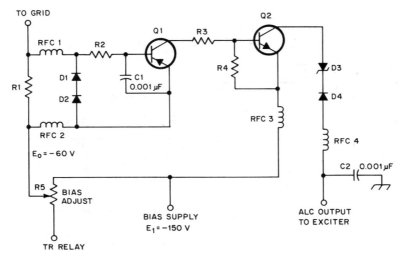

SEE PARTS LIST AND TEXT
FOR COMPONENT VALUES NOT SHOWN

Automatic Level Control for SSB Transmitters. Audio-derived ALC circuit allows maximum output from a SSB transmitter without "flat-topping" and spurious emissions. Circuit operates only on the DC currents in the grid circuit of the final amplifier of a SSB transmitter; thus, the circuit works in all modes (including "tune-up") and prevents damage to the final amplifier tube. Circuit is "universal" in design and component values need to be changed only in response to output power of the final amplifier stage and the frequency range (HF vs. VHF/UHF) over which the transmitter operates. Circuit shown was designed for use with a transmitter using a 4CX1000A for 1000 W of power from 1.8–54 MHz. D1, D2, and D4 are 1N4007 diodes or equivalent. D3 must have a voltage rating of approximately 20-V

less than the voltage at the emitter of Q2. In circuit shown, D3 is a 130-V, 0.5-W Zener diode. Q1 is a MPS-A92 or any PNP transistor with a 200-V or higher rating while Q2 is a MPS-A42 or other NPN transistor with a 200-V or higher rating; Q1 and Q2 are noncritical so long as they have a voltage rating above the highest voltage in the circuit. R1 is 4.7 K, R2 is 470 ohms, R4 is 100 K, and R5 is the existing transmitter bias control. The value of R3 is numerically equal to the difference in standby and operating biases; in circuit shown, it is 100 K. RFC1–4 were 1-mH chokes; use lower values for VHF/UHF operation. Original article includes more details on circuit adaptations and installation.—M. Mandelkern, "ALC for Class AB1 Amplifiers," QST, July 1986, pp. 38–39, 47.

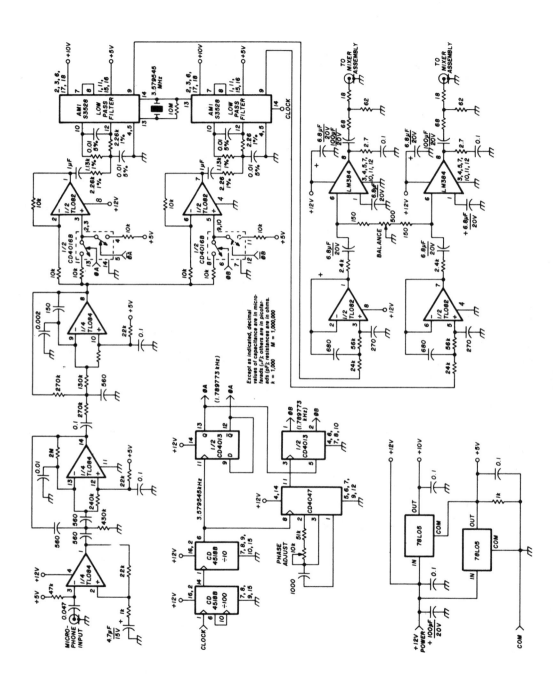

216

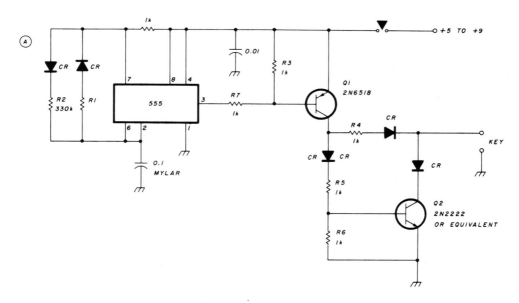

Tuning "Pulser" for SSB Amplifier Adjustment. Used to properly tune SSB linear amplifiers to prevent "splatter" and similar undesired emissions. Circuit produces maximum-peak plate current with a human voice duty cycle (about 33%) in a SSB transmitter. Circuit is used by placing the SSB transmitter in the CW mode, turning on the pulser, and advancing the carrier control until some ALC action is observed. Then the system's linear amplifier is switched on and the plate voltage is set for SSB. The pulser is keyed again and the amplifier is tuned for maximum relative output on the late-tune and load controls. All diodes labeled CR are silicon with 80-PIV ratings or higher. R1 can be either 160 K or 180 K; the higher value lengthens the output pulse. Circuit may be used with positive or negative keying systems.—R. Measures, "Adjusting SSB Amplifiers," *ham radio*, September 1985, pp. 33–36.

◀ **SSB Audio Modulator Using the "Weaver Technique."** Using the so-called "Weaver method" (after its developer, D. K. Weaver) by generating a SSB signal without crystal filters of accurate phasing and balancing of the modulator. Most of the circuitry operates at audio frequencies. Microphone input is applied to a wideband gain stage and then to a high pass filter in cascade with a low pass filter. The signal is then split into two paths, with each path consisting of a double-balanced mixer, a sharp low pass filter, a buffer stage, and 50-ohm pad designed to deliver 0-dBm to the mixer stages. Switched-capacitor filters are used instead of LC designs. A 1.8-kHz quadrature local oscillator and a pair of amplifiers for driving RF mixers complete the circuit.—N. Bernstein, "Two Meter Transmitter Uses Weaver Modulation," *ham radio*, July 1985, pp. 12–19.

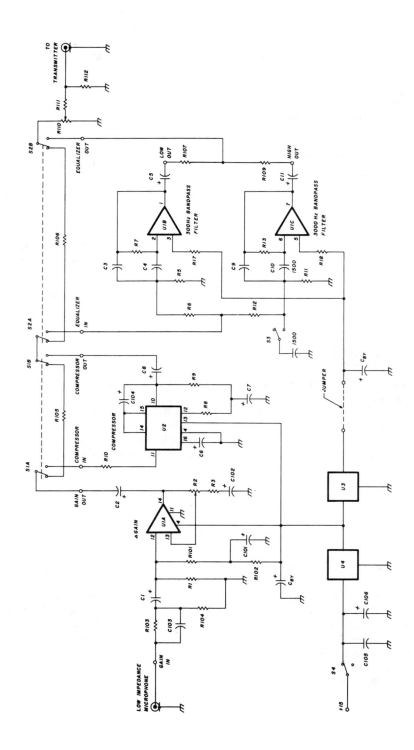

table 1. ACSSB Level One transmit adapter parts list.

item	description
	capacitors
C1	1 μF tantalum (TM 1/35)*
C2	1 μF, tantalum (TM 1/35)*
† C3	1000 pF Epoxy Dipped Ceramic (EDC) (272-154)**
† C4	1000 pF EDC (272-154)**
C5	1 μF, tantalum (TM 1/35)*
† C6	1 μF, tantalum (TM 1/35)*
C7	10 μF, tantalum (TM 1/35)*
C8	1 μF, tantalum (TM 1/35)*
† C9	1000 pF EDC (272-154)**
† C10	1000 pF EDC (272-154)**
C11	1 μF, tantalum (TM 1/35)*
C101	1 μF, tantalum (TM 1/35)*
C102	22 μF, tantalum (TM 22/6)*
C103	0.22 μF (TM 0.22/35)*
C104	1 μF, tantalum (TM 1/35)*
C105	0.1 μF(TM 0.1/35)* (272-158)**
C106	10 μF, tantalum (TM 10/25)*
CBY	1 μF, tantalum (TM 1/35)*
	integrated circuits
U1	MC3403 or equivalent
U2	NE570 or 571
	resistors
R1	47 kilohm
R2	10 kilohm potentiometer
R3	1 kilohm
R4	not used
† R5,R6	470 kilohm
† R7	1 Megohm
R8,R9	10 kilohm
R10	36 kilohm

item	description
† R11	470 kilohm
† R12	18 kilohm
† R13	180 kilohm
R14-R16	not used
R17,R18	470 kilohm
R103	10 kilohm
R102	47 kilohm
R101	680 kilohm
R104	470 kilohm
R105	1 kilohm (see note)
R106	1 kilohm (see note)
R107	1.5 kilohm
R108	3.3 kilohm
R109	4.7 kilohm
R110	1 kilohm potentiometer
R111	150 kilohm
R112	470 kilohm
	switches
S1,2	DPDT (275-1546)**
S3,4	SPDT (275-625 or 647)**
	voltage regulators
U3	7806 or 78L05
U4	7812 or 78L12
	miscellaneous
box, lid	Pomona 2902
knob	to fit R2 potentiometer
connectors as required	
standoffs 4-40 threaded × 1 inch (25.4 mm) long	

Note: Select for about 3-6 dB loss through adapter in the out position with ΔG = 0 and drive potentiometer = max.

*   JAMECO
** Radio Shack
†   5 percent tolerance, if possible

ACSSB System-Adapter for Reception and Transmission. Separate transmitting and receiving adapters designed for amateur radio stations experimenting with "amplitude companded SSB," or ACSSB, communications. ACSSB basically involves compression of the signal amplitude at transmission and its expansion on the receiving end. This permits narrower bandwidths than conventional SSB, more effective use of transmitter power, and a greatly improved signal-to-noise ratio on reception. Circuits shown offer a 2:1 compression ratio on transmit, a 1:2 expansion ratio on receive, and spectrum equalization on both transmission and reception. Unlike commercial systems, however, no pilot-tone reference is included to aid in tuning the received signal; this must be done manually. Table 1 shows the parts used by the transmit adapter while table 2 shows the parts used by the receiving adapter (pages 220–221). Original article includes extensive details on ACSSB theory, circuit construction and PC board layout, and proper tuning of transmitters and receivers for ACSSB communication.—J. Eagleson, "ACSSB: A Level-One Adapter," *ham radio*, October 1986, pp. 10–28.

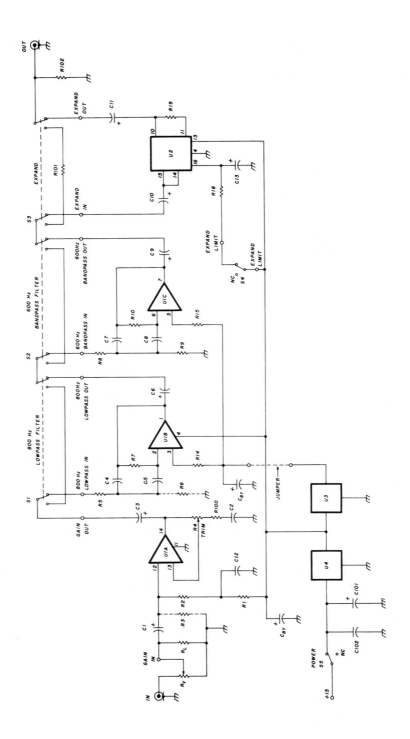

table 2. ACSSB Level One receive adapter parts list.

item	description
	capacitors
C1	1 μF (TM 1/35)*
C2	22 μF (TM 22/6)*
C3	1 μF (TM 1/35)*
† C4	4700 pF 10 percent (MY 0.0047/100)* (272-155)**
† C5	0.1 μF 10 percent (MY 0.1/100)* (272-158)**
C6	1 μF (TM 1/35)*
† C7,C8	0.01 μF 10 percent (MY 0.01/100)*(272-156)**
C9,C10,C11	1 μF (TM 1/35)*
C12	4.7 μF (TM 4.7/35)*
† C13	C13 1 μF 5 percent (TM 1/35 OK)*
CBY	1μF (TM 1/35)*
C101	10 μF (TM 1/35)*
C102	0.1 μF (MY 0.1/100)
	resistors
R1	10 kilohm
R2,R3	47 kilohm
R4	10 kilohm trim potentiometer
† R5	36 kilohm
R6	(not used)
† R7	36 kilohm
† R8	18 kilohm
R9	47 kilohm
† R10	36 kilohm
R11-R13	(not used)
R14,R15	47 kilohm
R16,R17	(not used)
R18	3.3 Megohm
R19	36 kilohm
RL	(load resistor to match source. 8 ohms or 100 kilohms, typical)
R100	1 kilohm (jumper if more gain is needed)
R101	4.7 kilohm
R102	10 kilohm
Rv	potentiometer, 500 ohms (low impedance in) 100 kilohm (high impedance in) (audio taper)
	switches
SW1-4	DPDT (275-1546)**
SW5	SPDT (275-625 or 647)**
	integrated circuits
U1	MC3403 or eqivalent*
U2	NE570 or NE571*
	voltage regulators
U3	7806 (78L05 OK)* **
U4	7812 (78L12 OK)* **

* JAMECO
** Radio Shack
† 5 percent tolerance, if possible

LINEAR AMPLIFIER

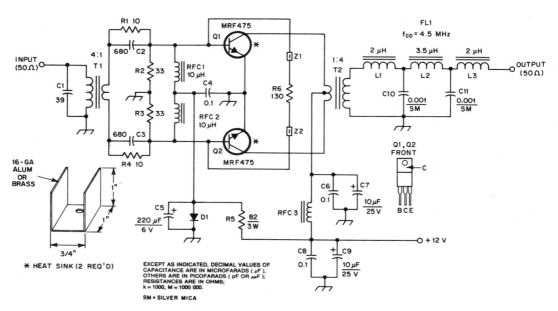

10-W Broadband RF Amplifier for SSB.

An input signal of 1–2 W will produce a 10-W output from 1.8–30 MHz when an appropriate filter is used at FL1. A pair of MRF475 transistors are used in a push-pull configuration. Broadband transformers (T1 and T2) are used to provide a 50-ohm impedance at input and output; C1 across the primary of T1 tunes out unwanted reactance in the 14–30 MHz range. Bias for class AB service is developed by D1. Efficiency of this circuit is between 50–60%. FL1 is a low pass filter designed for a cutoff frequency slightly above the highest desired operating frequency; in circuit shown, this is 4.5 MHz. Filter constants and parts values for other frequencies can be obtained using standard design formulas. TO-220 transistors may be substituted for the MRF475 units.

D1 is a 2-A, 50-PIV silicon rectifier. L1 and L3 are 20 turns of #22 wire on a T50-2 toroid core. L2 is 26 turns of #24 wire on a T50-2 toroid core. RFC3 is five turns of #22 wire on a FT50-43 form. T1's form consists of two rows of three FT37-43 toroids; toroids are glued together to form two sleeves and then sleeves are glued together side by side to form balun core. T1's primary is four turns of #24 wire while the secondary is two turns of small insulated "hookup" wire. T2 has a primary of one turn of #22 "hookup" wire on a #43 ferrite balun core with a secondary of two turns of #22 "hookup" wire. Z1 and Z2 are #43 ferrite beads.—D. DeMaw, "The Principles and Building of SSB gear," *QST*, January 1986, pp. 29–32.

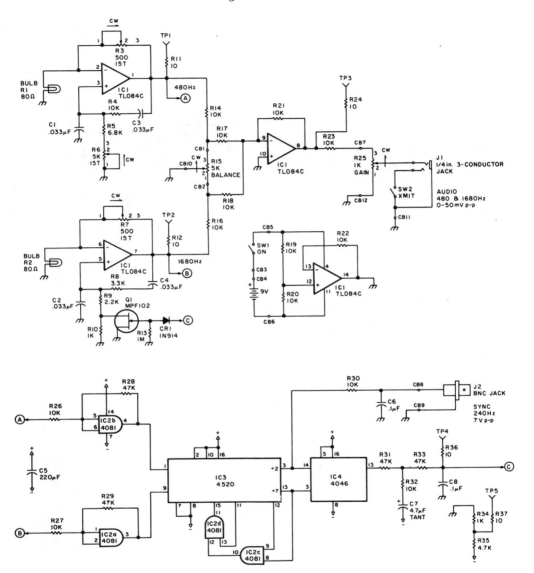

Two-Tone Test Unit for SSB Transmitters.   Used in conjunction with an oscilloscope to generate RF patterns produced by two modulating tones applied to a SSB transmitter. Circuit generates two low distortion audio tones in the 300–2100 Hz range, with each tone at least 1000 Hz apart in frequency using a PLL-controlled oscillator. A "synch output" for the oscilloscope is also provided. Dual oscillators operating at 480 and 1680 Hz are used to generate output signals, and the circuit is built in separate analog and digital sections. Original article includes information on constructing the unit and running two-tone tests.—F. Perkins, "Build the Dixie Whistler," *73 Magazine*, April 1985, pp. 36–44.

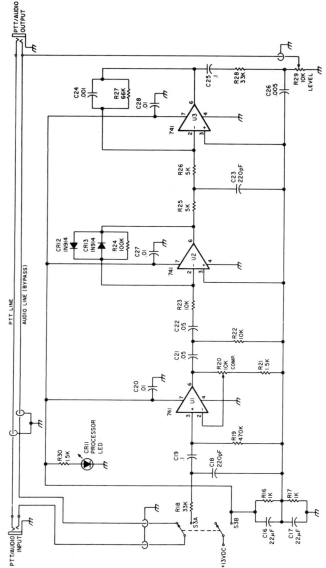

SSB Speech Processor. Processes audio input to deliver a more constant and effective output from a SSB transmitter. The first stage is a common audio amplifier whose gain is controlled by potentiometer R20. The second op amp is an audio amplifier with diodes in the feedback loop. If U1's output remains below ± 0.6 V DC, CR12 and CR13 will conduct and limit the gain of U2. This action keeps the audio level constant on U2's output. U3 is a bandpass filter removing unwanted harmonics and R29 controls the output level. Three 741 op amps were used instead of one quad op amp for easier placement. Circuit was originally designed for use with an "Argonaut" amateur radio transceiver but can be easily adapted for use with similar units.—F. Marcellino, "Get Organized Today with Argo's Helper," 73 *Magazine,* March 1986, pp. 44–49.

224

SPEECH AMPLIFIER

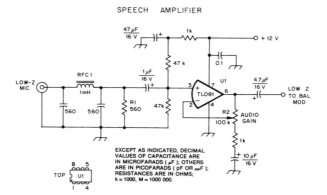

SSB Speech Amplifier. Amplifies speech input from a microphone to a level sufficient to drive the balanced modulator of a SSB transmitter. Input impedance is approximately 500 ohms. If a higher impedance microphone is being used, delete resistor R1 (560 ohms). The TL081 delivers a gain of 40 dB. Input frequencies below 200 Hz are suppressed; excessive high frequencies can be rolled off to ground by adding bypass capacitors between ground and pins 3 and 6 of the TL081. Values between 0.01–0.47 μF are suggested.—D. DeMaw, "The Principles and Building of SSB Gear." QST, October 1985, pp. 27–30.

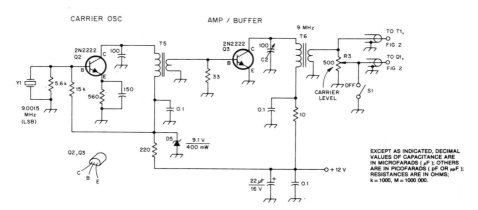

Carrier Generator for SSB Transmitters. Generates a 9-MHz signal for processing in a balanced modulator. The 9.0015-MHz crystal produces a LSB signal; a 8.9985-crystal can be substituted if a USB signal is desired. A regulated operating voltage for Q2 is provided by Zener diode D5. T5 is a narrowband RF transformer whose primary is 24 turns of #26 wire on a T50-2 toroid core and a secondary of five turns of #26. T6's primary is 28 turns of #26 wire on a T50-2 powdered iron core and its secondary is eight turns of #26 wire. The frequency of crystal Y1 should fall 20 dB below the center frequency of the SSB transmitter's bandpass filter; a 60-pF trimmer capacitor can be added between the lower end of Y1 and ground to adjust the circuit's output for most natural audio.—D. DeMaw, "The Principles and Building of SSB Gear," QST, October 1985, pp. 27–30.

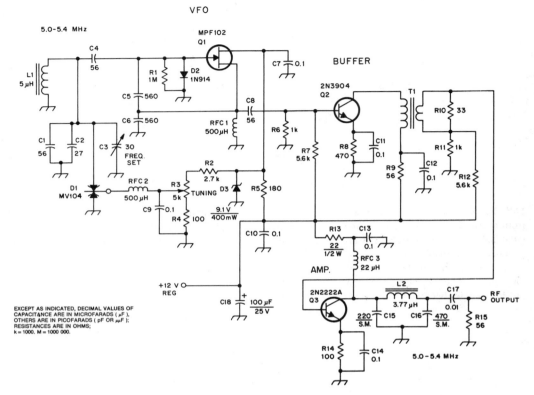

EXCEPT AS INDICATED, DECIMAL VALUES OF
CAPACITANCE ARE IN MICROFARADS (μF),
OTHERS ARE IN PICOFARADS (pF OR μμF);
RESISTANCES ARE IN OHMS;
k = 1000. M = 1000 000.

5-MHz VFO. Circuit incorporates buffer and amplifier stages and gives output from 5–5.25-MHz. Voltage-variable capacitor diode tuning, using a Motorola MV104 or equivalent, is used instead of a bulky, expensive air-variable capacitor. The MPF102 is a JFET device stabilized by D2 (1N914). The 2N3904 acts as a class A buffer amplifier. Output amplifier Q3 (2N2222A) is also operated as class A. Circuit output impedance is 50 ohms. L1 is 32 turns of #24 wire on a T68-6 toroidal core. L2 is 30 turns of #30 wire on a T-37-2 toroidal core. T1 has a primary of 15 turns of #30 on a FT37-43 toroid and a secondary of four turns of #30. Zero-temperature-coefficient (NPO) capacitors should be used for C1–C6 and C8.—D. DeMaw, "The Principles and Building of SSB Gear," *QST*, December 1985, pp. 37–40.

Crystal-Controlled Local Oscillator for SSB Transmitters. Provides a 5–5.25-MHz output signal at a single frequency for the mixer stage of a SSB transmitter. For maximum frequency stability, components with low temperature coefficients should be selected.—D. DeMaw, "The Principles and Building of SSB Gear," *QST*, December 1985, pp. 37–40.

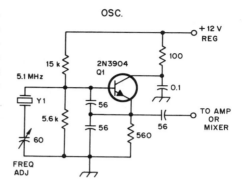

OSC.

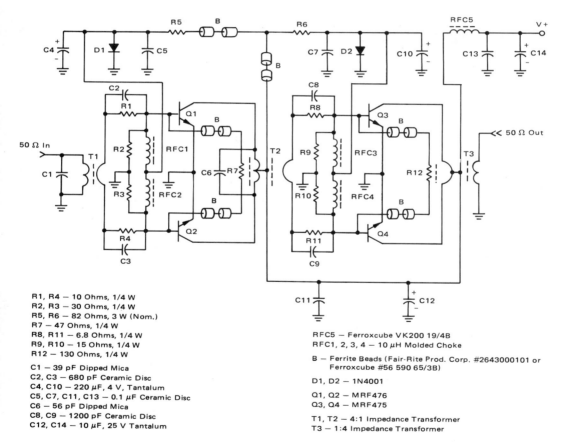

R1, R4 — 10 Ohms, 1/4 W
R2, R3 — 30 Ohms, 1/4 W
R5, R6 — 82 Ohms, 3 W (Nom.)
R7 — 47 Ohms, 1/4 W
R8, R11 — 6.8 Ohms, 1/4 W
R9, R10 — 15 Ohms, 1/4 W
R12 — 130 Ohms, 1/4 W

C1 — 39 pF Dipped Mica
C2, C3 — 680 pF Ceramic Disc
C4, C10 — 220 μF, 4 V, Tantalum
C5, C7, C11, C13 — 0.1 μF Ceramic Disc
C6 — 56 pF Dipped Mica
C8, C9 — 1200 pF Ceramic Disc
C12, C14 — 10 μF, 25 V Tantalum

RFC5 — Ferroxcube V K200 19/4B
RFC1, 2, 3, 4 — 10 μH Molded Choke

B — Ferrite Beads (Fair-Rite Prod. Corp. #2643000101 or Ferroxcube #56 590 65/3B)

D1, D2 — 1N4001

Q1, Q2 — MRF476
Q3, Q4 — MRF475

T1, T2 — 4:1 Impedance Transformer
T3 — 1:4 Impedance Transformer

Broadband SSB Driver. Designed to operate from 1.6–30 MHz and deliver 25-dB gain with a maximum output of 20 W. Distortion is low, with a typical IMD of better than −35 dB. Class A operation is used to achieve this figure. Original applications note gives construction data and PC-board artwork.—"Low Distortion 1.6 to 30 MHz SSB Driver Designs," AN-779, Motorola, Inc.

23
Telephone Circuits

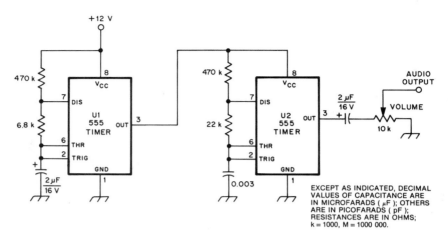

"Beeper" Generator. Produces a "beep" tone every 3–4 seconds when power is applied. Output can be fed into other audio circuits for such purposes as indicating when a telephone call is being recorded or if a transmission is being relayed. The 10-K potentiometer at the circuit's output allows the level of the "beep" tone to be set so that it can be heard but not excessively interfere with the communication.—B. Heil, "A VHF/UHF Remote Base Station," *QST*, July 1986, pp. 30–33.

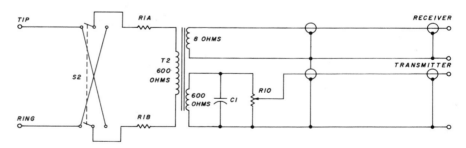

Phone Patch. Simple circuit allows interfacing amateur radio station to telephone systems operating on the American standard. C1 is 0.1 μF. R1A and R1B are 100–270 ohms; the proper value for best audio quality will depend upon the characteristics of the telephone line and the station equipment. Potentiometer R10 should be 1 K and is adjusted for desired sound level.—J. Macassey, "Building and Using Phone Patches," *ham radio*, October 1985, pp. 34–38.

Dialing-Tone Decoder Using PLL. Circuit shown uses seven NE567 PLLs which have been designed specifically with tone-decoding applications in mind. Circuit accepts standard dialing-tone inputs to produce indicated outputs. Bandwidth is approximately 8% at 100-mV input and 5% at 50-mV input. Typical component values are shown in the lower right.— "Selected Circuits Using the NE567," AN188, Signetics Corporation.

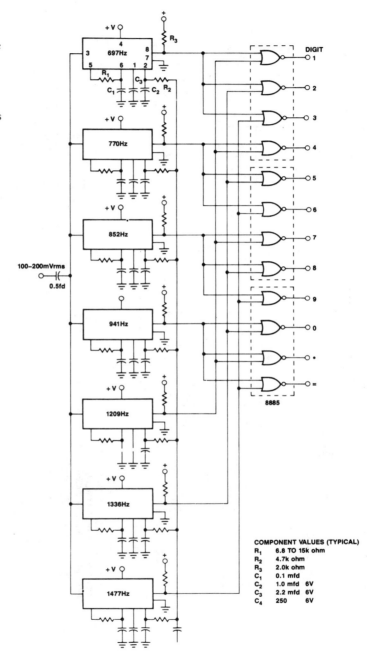

COMPONENT VALUES (TYPICAL)

R_1	6.8 TO 15k ohm
R_2	4.7k ohm
R_3	2.0k ohm
C_1	0.1 mfd
C_2	1.0 mfd 6V
C_3	2.2 mfd 6V
C_4	250 6V

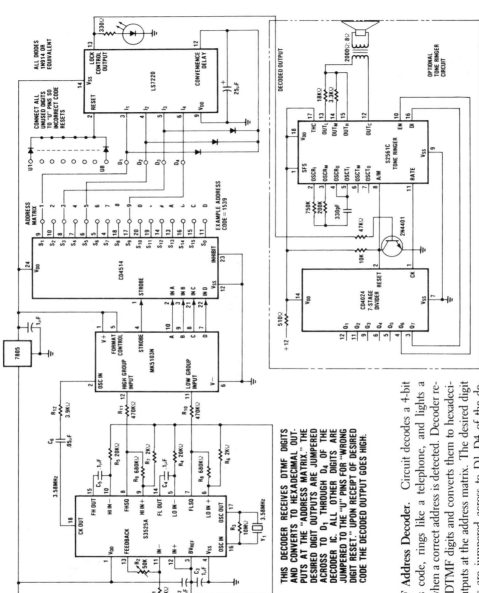

DTMF Address Decoder. Circuit decodes a 4-bit address code, rings like a telephone, and lights a LED when a correct address is detected. Decoder receives DTMF digits and converts them to hexadecimal outputs at the address matrix. The desired digit outputs are jumpered across to D1–D4 of the decoder IC. All other digits are jumpered to the "U" pins to indicate wrong-digit reset. Upon receipt of the proper code, the decoded output goes high. Circuit can be modified to initiate remote control func-

tions.—"Using the S3535A/B DTMF Bandsplit Filter," Applications Note, Gould/AMI Semiconductors.

THIS DECODER RECEIVES DTMF DIGITS AND CONVERTS TO HEXADECIMAL OUTPUTS AT THE "ADDRESS MATRIX." THE DESIRED DIGIT OUTPUTS ARE JUMPERED ACROSS TO D₁ THROUGH D₄ OF THE DECODER IC. ALL OTHER DIGITS ARE JUMPERED TO THE "U" PINS FOR "WRONG DIGIT RESET." UPON RECEIPT OF DESIRED CODE THE DECODED OUTPUT GOES HIGH.

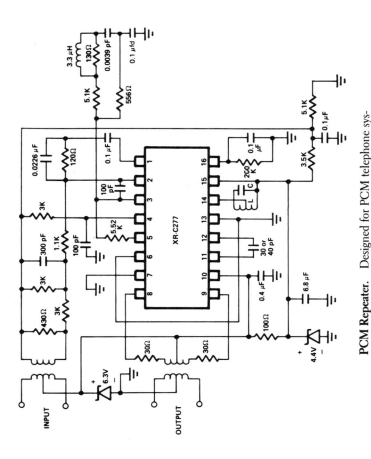

PCM Repeater. Designed for PCM telephone systems. Inductor L and capacitor C are set to form a high Q tank resonant at 1.544 MHz for a 1.544-MPS data rate T-1 repeater system. XR-C277 is a low voltage PCM repeater IC.—"XR-C277 Low Voltage PCM Repeater IC," AN-04, Exar Corporation.

Dial-Tone Detector. Indicates presence or absence of a dial tone. Precision dial tone is a combination of 350 Hz and 440 Hz. By using a crystal of 1.758 MHz with the S3525 IC, the 3-dB points of the low-group filter output will be 334–496 Hz. Thus, all the energy from precision dial tone will be available at the low-group output. Any signal present at the high-group output will indicate the presence of something other than a dial tone.—"Using the S3525A/B DTMF Bandsplit Filter," Applications Note, Gould/AMI Semiconductor.

233

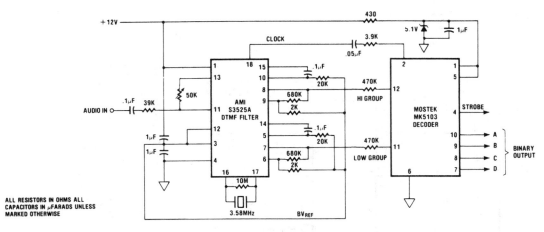

DTMF Decoder. S3525A bandsplit filter and MK5103 DTMF decoder together produce a 4-bit binary output from DTMF audio input. Standard 3.58 crystal is used as time base for the S3525A, which then provides a buffered time-base output for the decoder.—"S3525A DTMF Bandsplit Filter," data sheet, Gould/AMI Semiconductor.

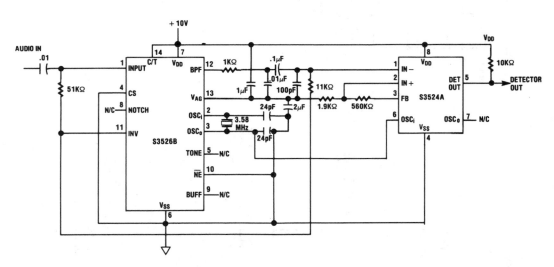

Digital Frequency-Detector for 2600 Hz. S3524A digital frequency-detector IC is capable of determining if an incoming audio tone is within 35 Hz, high or low, of 2600 Hz. Input is first processed by S3526B 2600-Hz bandpass filter. Every period of incoming signal is checked, and a true output is given for each period falling within the desired bandwidth.—"S3524A 2600 Hz Digital Frequency Detector," data sheet, Gould/AMI Semiconductors.

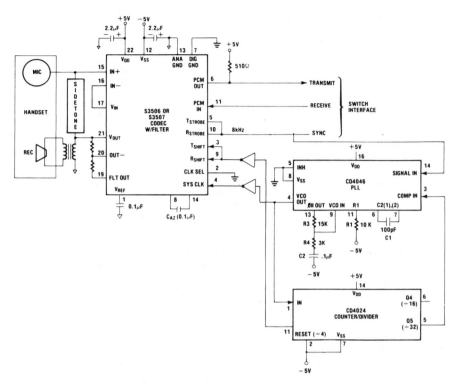

Digital Telephone System. Implements PCM technique for voice data in telephone systems. The S3506-3507 is a single-chip codec which serves as the heart of this system. Since asynchronous operation is not necessary, common transmit-and-receive timing signals are used. A 4046 PLL IC derives the 2048-kHz system clock and 64-kHz shift clock from the 8-kHz synchronizing signal received from the switch. The synchronizing signal also serves as the transmit/receive strobe signal; its duty cycle is not important for codec operation. Microphone output feeds directly into the coder input while the decoder output drives the receiver through an impedance transformer.—"S3506/S3507/S3507A CMOS Single Chip μ-Law/A-law Synchronous Combo Codecs with Filters," data sheet, Gould/AMI Semiconductors.

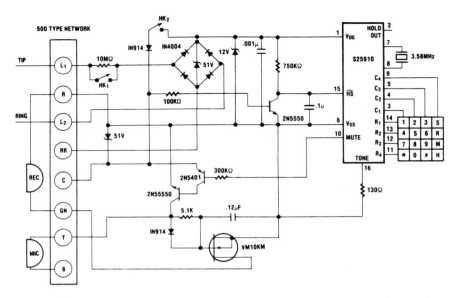

Ten-Memory DTMF Dialer. S25910 IC stores up to ten 16-bit numbers and also has a 16-digit redial buffer. Circuit also offers normal DTMF dialing functions. "M" key dials a number in memory, "H" key places caller on hold, "R" redials last number called, and "S" enters a number into memory. Supply voltage may range from 2.5–10 V.—"S25910/S25912 Ten Memory DTMF Dialer," data sheet, Gould/AMI Semiconductors.

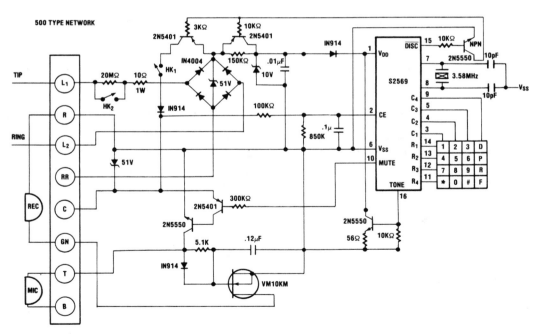

Line-Powered DTMF Dialer with Redial. Uses Gould/AMI S2569 DTMF tone generator IC as heart of system. Separate keys initiate the disconnect (key "D"), pause (key "P"), redial (key "R"), and flash (key "F") functions. Supply voltage can range from 2.5–10 V.—"S2569/S2659A DTMF Tone Generator with Redial," data sheet, Gould/AMI Semiconductors.

24

Television and Video Circuits

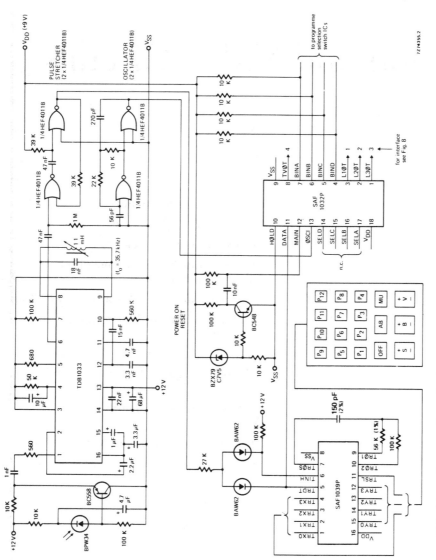

Television Remote Control System. Uses a SAF1032P (receiver/decoder) and SAF1039P (transmitter) pair to form a remote control system using pulse-code modulation. Transmitter frequency is 120 kHz and 16 selection codes are available. The receiver responds to 32 different control commands.

Original article includes an operating code and function table. System may be adapted for industrial equipment and similar applications.—"SAF1032P/39P R/C Receiver; R/C Transmitter," 1985 Linear LSI Data and Applications Manual, 5-107, Signetics Corporation.

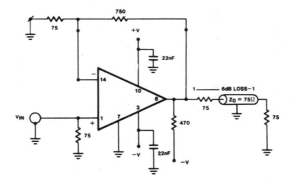

Color Video Amplifier. NE5539 wideband operational amplifier is easily adapted for use as a color video amplifier. Circuit is designed for a 75-ohm input and output impedances and offers a gain of 20 dB for 8-V supply voltages. Linearity across the video-signal bandwidth is typically 0.5% or better.—"SE/NE5539 Ultra High Frequency Operational Amplifier," 1985 Linear LSI Data and Applications Manual, 6-123, Signetics Corporation.

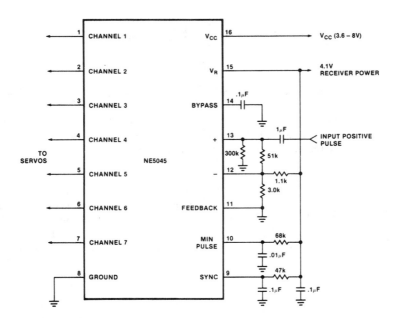

Remote Control Decoder. NE5045 decoder IC produces seven parallel outputs based upon a serial input. Pulse-width modulation is used, and signals as low as 10 mV peak-to-peak can be detected. The NE5045's internal voltage regulator provides power supply regulation for the decoder and for a radio receiver if desired. Each output is capable of sourcing 2 mA and sinking 1 mA of current.—"NE5045 Seven Channel RC Decoder," 1985 Linear LSI Data and Applications Manual, 7-13, Signetics Corporation.

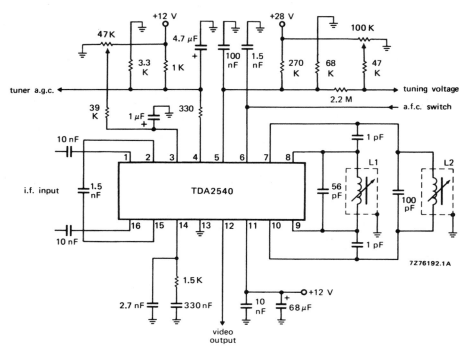

Video IF Amplifier and Demodulator with AGC and AFC. Circuit operates at 38.9 MHz and delivers a 2.7-V output signal for a 100-μV input level. Bandwidth is 6 MHz. The IF gain control range is typically 64 dB with a signal-to-noise ratio of 58 dB.

The AFC function can be switched on or off by a DC voltage. Q factor of L1 and L2 is approximately 80.—"TDA2540 Video IF/AFT," 1985 Linear LSI Data and Applications Manual, 8-56, Signetics Corporation.

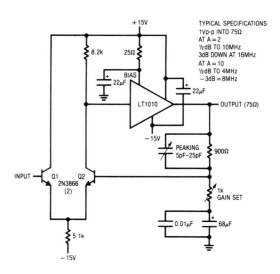

Adjustable-Gain Video Amplifier. Drives 1 V peak-to-peak into a 75-ohm load. The differential pair (2N3866) provides gain with the LT1010 op amp serving as an output stage. Feedback is arranged in the conventional manner, although the 0.01- and 68-μF capacitor pair limits DC gain to unity for all gain settings. For applications sensitive to NTSC requirements, dropping the 25-ohm outstage bias-value will aid performance.—"LT1010 Fast ±150 mA Power Buffer," data sheet, Linear Technology Corporation.

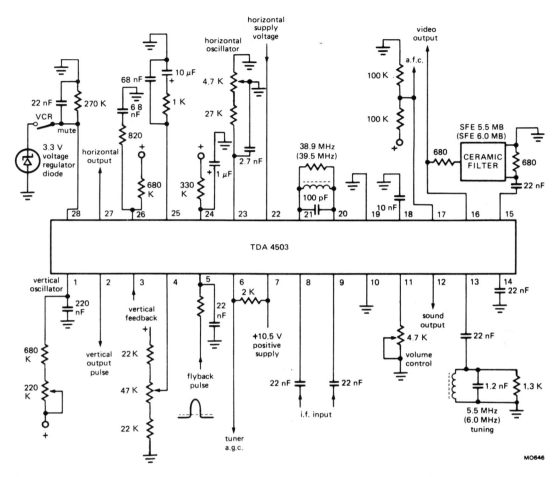

Small-Signal Strip for Monochrome Television. TDA4053 combines all small-signal functions, except tuning, required for monochrome television reception. For complete receiver, the only output stages required are horizontal and vertical deflection, video, and sound. Among functions provided are horizontal and vertical output, synchronous demodulation, audio output, IF amplification, AFC and AGC detection, and sync separation.—"TDA4503 Small Signal Combination for Monochrome TV," 1985 Linear LSI Data and Applications Manual, 8-164, Signetics Corporation.

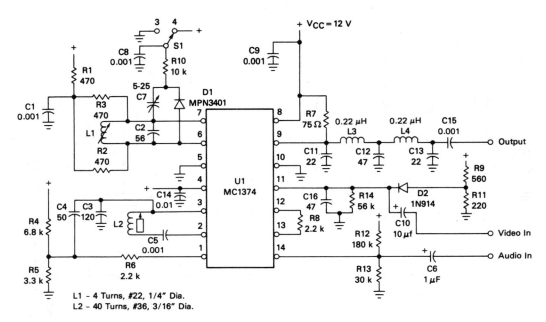

L1 - 4 Turns, #22, 1/4" Dia.
L2 - 40 Turns, #36, 3/16" Dia.

Television Modulator.   Produces a video signal on channels 3 or 4 from video and audio inputs; circuit was originally intended for use with video cassette recorders, subscription television decoders, video games, etc. Output is about 12-mV RMS. This is 12 dB greater than permitted by FCC rules, so it must be "padded down" for commercial applications. Original article includes PC-board artwork and a parts placement diagram.—"Application of the MC1374 TV Modulator," AN-829, Motorola, Inc.

5-W Audio Stage for Television Receivers. TDA2611A audio amplifier IC provides 5 W into an 8-ohm speaker. Frequency response exceeds 15 kHz with total harmonic distortion of less than 1%. The supply voltage can range from 6–35 V. —"TDA2611A 5W Audio Output for TV," 1985 Linear Data and Applications Manual Volume II, 6-70, Signetics Corporation.

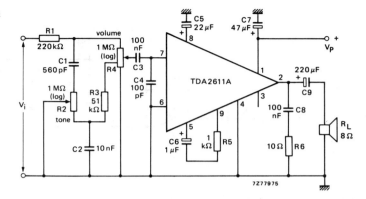

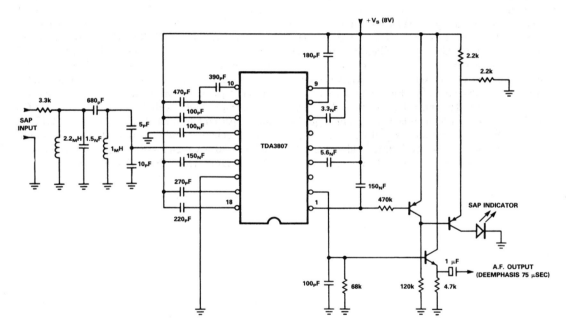

Multi-Channel Television Sound Decoder. TDA3807 provides amplification, demodulation, indication, and no-signal muting for processing multi-channel television audio. Audio bandwidth is 9 kHz with a signal-to-noise ratio of 64 dB.—"TDA3807 Second Audio Program," 1985 Linear Data and Applications Manual, Volume II, 5-69, Signetics Corporation.

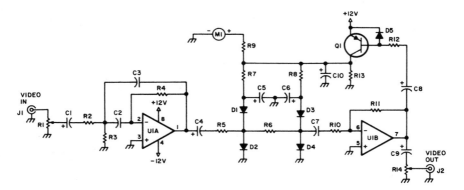

Satellite Video-Prefilter and AGC. Helps maintain the level of a video signal received from a weather satellite using the WEFAX standard, which uses amplitude-modulated tones of around 2400 Hz. This circuit also adds some prefiltering of the signal and is placed between the video source (receiver, tape, etc.) and video display system. U1A is configured as a unity-gain active filter with a center frequency of 2400 Hz and a bandwidth of 1600 Hz. U1B is configured to maintain a dynamic equilibrium in the face of dynamic signal levels. Potentiometer R1 controls the AGC threshold while R14 controls the circuit gain. U1A/B are halves of a MC1458 dual op amp and Q1 is a 2N4403 or equivalent general purpose PNP audio transistor. D1 through D5 are 1N914 diodes. Potentiometers R1 and R14 are both 10 K. R2, R7, and R8 are 10 K, while R5 and R6 are 4700 ohms. R9 and R13 are 47 K and R11 and R12 are 120 K. R3 is 2700 ohms, R4 is 20 K, and R10 is 1000 ohms. C1, C4, C8, and C9 are all 1-μF tantalum capacitors. C2, C3, and C7 are 0.01-μF Mylar with 100-V ratings. C5 and C6 are 22-μF tantalum rated at 6 V. C11 and C12 are 0.1-μF ceramics and C10 is a 100 μF aluminum. M1 is a 50-μA meter.—R. Taggert, "Weathersat," 73 *Magazine*, December 1986, pp. 58–60.

25

Test and Measurement Device Circuits

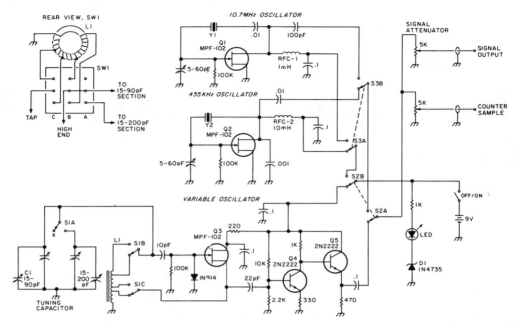

3–22-MHz RF Signal Generator.   Covers 3.2–22 MHz in two bands and produces a 4-V output into a 500-ohm load. Two outputs are provided, with one providing a signal sample for a frequency counter. Output strength is controlled by a 5-K potentiometer. Zener diode D1 (1N4735) and LED D2 make a simple ON/OFF and battery-status display; when the battery voltage falls below the threshold set by D1, D2 no longer lights. Y1 is a 10.7-MHz crystal, while Y2 is a 455-kHz ceramic resonator. L1 is 38 turns of #28 wire on a T50-2 form with taps at four, 10, and 17 turns.—R. Littlefield, "Conjure an RF Genie," 73 *Magazine*, November 1985, pp. 32–33.

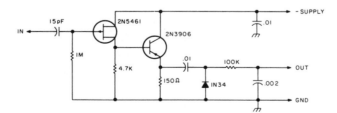

RF Probe.   Can detect RF signals from 15 kHz to beyond 30 MHz at levels up to 2.5 V. Although designed to work with a −5 V DC supply, probe can also be used with supplies up to −15 V DC. DC output voltage of probe is about 20% lower than the RMS input.—R. Bailey, "Probe the Unknown," 73 *Magazine*, November 1985, pp. 46–47.

Inductance Meter. Used in conjunction with a frequency counter, circuit provides unknown inductance not only with a selectable tuning capacitor but also with an adjustable negative resistance to make it oscillate. Inductance can be measured from small fractions of a microhenry to thousands of henrys over a frequency range of a few Hz to tens of MHz. Potentiometer R1 should be adjusted so that the circuit barely oscillates. Original article gives details on using the meter.—R. Ketchledge, "The Incredible Inducto-Gauge," 73 *Magazine,* July 1985, pp. 34–36.

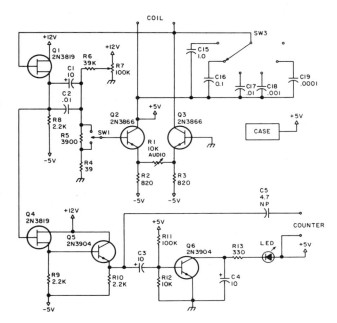

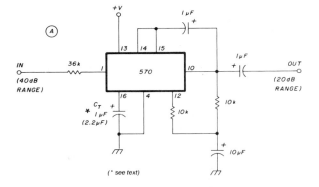

VU-Meter Range Compressor and Expander. Circuits compress or expand range of a VU meter by a 2:1 ratio, as indicated by strengths of the input and output signals. Substituting a 2.2-μF capacitor for the 1-μF capacitor, as indicated in both schematics, will extend the performance range of both circuits down to 100 Hz at the cost of some degraded performance at higher frequencies.—J. Eagleson, "Extended Range VU Meter," *ham radio,* September 1986, pp. 59–61.

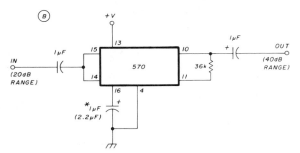

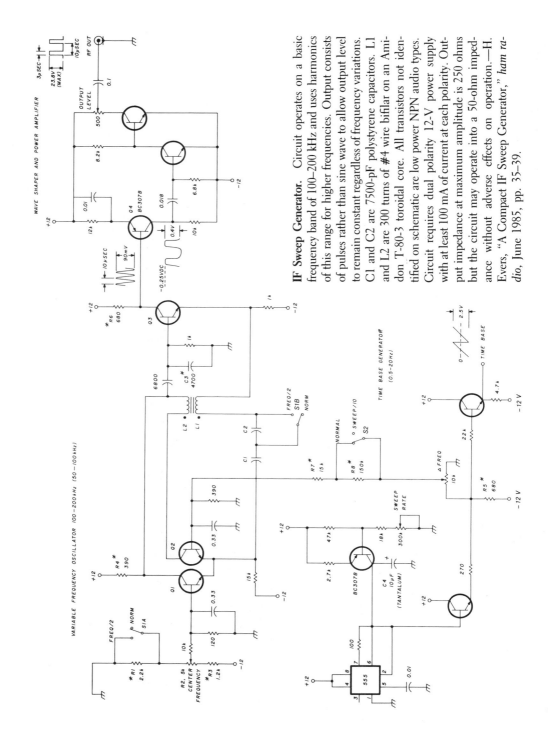

IF Sweep Generator. Circuit operates on a basic frequency band of 100–200 kHz and uses harmonics of this range for higher frequencies. Output consists of pulses rather than sine wave to allow output level to remain constant regardless of frequency variations. C1 and C2 are 7500-pF polystyrene capacitors. L1 and L2 are 300 turns of #4 wire bifilar on an Amidon T-80-3 toroidal core. All transistors not identified on schematic are low power NPN audio types. Circuit requires dual polarity 12-V power supply with at least 100 mA of current at each polarity. Output impedance at maximum amplitude is 250 ohms but the circuit may operate into a 50-ohm impedance without adverse effects on operation.—H. Evers, "A Compact IF Sweep Generator," *ham radio*, June 1985, pp. 35–39.

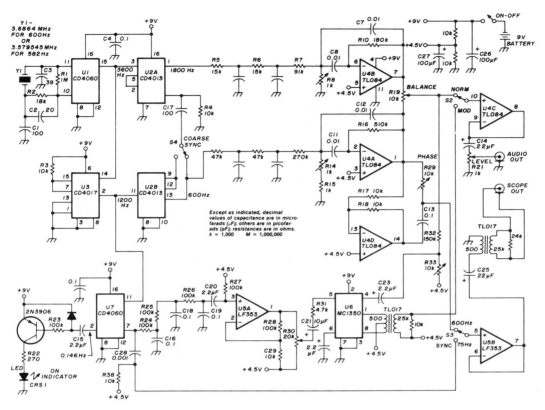

Two-Tone Test Generator. Generates two, low distortion audio tones at 75 Hz and 600 Hz for testing SSB transmitters. Output of transmitter modulated by this circuit is displayed on an oscilloscope, where linearity shows up as a straight line which is easier to judge than interlaced sine waves. Circuit also modulates the two tones at a rate similar to that of voice, permitting operation at average-voice power levels while a two-tone test is displayed. Circuit is crystal controlled, which permits accurate frequency-measurement of a transmitter without the necessity of generating a carrier. Original article includes information on proper use of the circuit in testing SSB transmitters.—B. McLagan, "Two-Tone Signal Generator," *ham radio*, February 1986, pp. 25–33.

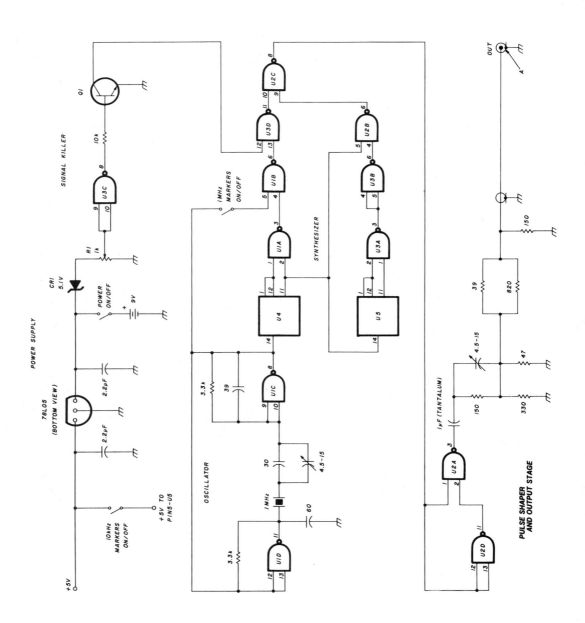

SWR and Watt Meter. Gives indication of transmitter power in forward and reverse directions when placed between transmitter output and the transmitting antenna. Circuit can handle up to 1000 W of power on frequencies between 1.8–30 MHz. D1 and D2 are 1N34A or similar diodes, while T1 is 60 turns of #30 wire wound on the T50-2 toroid core. T1 surrounds a length of 50-ohm coaxial cable; type RG-8X is recommended. Original article gives details on construction and calibration.—D. DeMaw, "The SWR Twins—QRP and QRO," *QST*, July 1986, pp. 34–37.

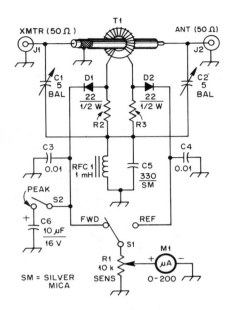

◀ **Frequency-Stable, High Level RF Signal Generator.** Entire circuit is contained in a closed, shielded metal box which is electrically connected to circuit ground only through the coaxial cable to the output connector at point A. This prevents stray signals from being present in the output. Circuit produces output pulses at 1 MHz, 100 kHz, and 10 kHz intervals which are selected by internal switches. U1 and U3 are 7400 quad NAND ICs, U2 is a 74S00, and U4 and U5 are 74LS90 decade counter ICs.—H. Evers, "A Frequency and Level Standard," *ham radio*, January 1986, pp. 10–20.

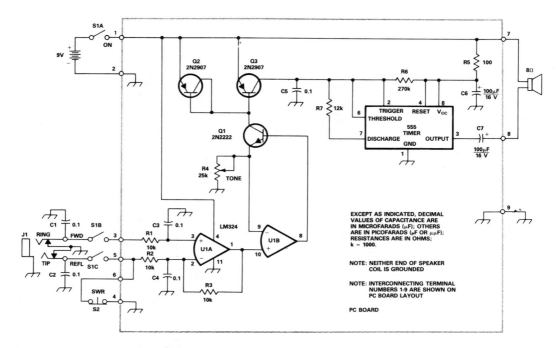

Audible RF Output and SWR Indicator. Produces one audible tone when output from a transmitter is detected and produces another tone if a reflected wave (indicating SWR) is detected. In use, transmitter is tuned to produce the highest possible tonal pitch from the speaker without losing the tone altogether. Pressing S2 will change the tone if SWR is present. The antenna matching network is adjusted while pressing S2 until both tones match, indicating that the SWR is then 1:1. U1 is a LM324 quad op amp functioning as a differential amplifier for the 555 timer, which operates as a voltage-to-audio frequency converter. Q1, Q2, and Q3 function as a current mirror, necessary for the unit to produce a usable range of audio tones. R4, the 25-K potentiometer, is adjusted once to provide usable tones. Unit requires samples of the forward and reflected DC voltages from the transmission line between transmitter and antenna matching network.—G. Murphy, "Meet the SWAILER," *QST*, January 1986, pp. 37–39.

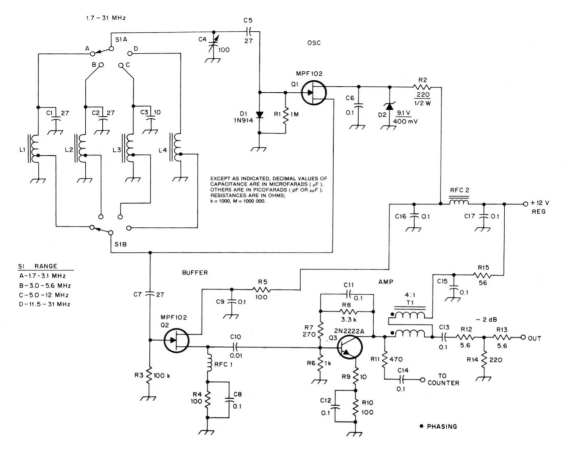

Tunable 1.7–31-MHz HF Signal Generator. Tapped-coil Colpitts-oscillator design provides tuning ranges of 1.7–3.1, 3–5.6, 5–12, and 11.5–31 MHz. Zener diode D2 is used at Q1 (MPF102) to lower the operating voltages of the oscillator. Q2 (another MPF102) is a source follower buffer stage. The source of Q2 is broadly tuned by RFC1. Q3 (2N2222A) is operated as a broadband class A amplifier; amplifier response is relatively flat from 1.5–35 MHz. C1, C2, C3, C5, and C7 should be NPO ceramic types. L1 is 32 turns of #26 wire on a FT-50-61 ferrite toroid tapped eight turns from ground. L2 is 18 turns of #24 wire on a FT-50-61 ferrite toroid tapped five turns from ground. L3 is 41 turns of #26 wire on a T50-2 toroid tapped 10 turns from ground. L4 is 21 turns of #26 wire on a T50-6 toroid. RFC2 is 15 turns of #26 on a FT-37-43 ferrite toroid. T1 is a broadband bifilar-wound transformer with a 4:1 impedance ratio; it uses 12 turns of #26 wire, twisted eight turns to the inch before winding, on a FT-50-43 ferrite toroid.—D. DeMaw, "Build a Homemade Signal Generator," *QST*, January 1986, pp. 40–43.

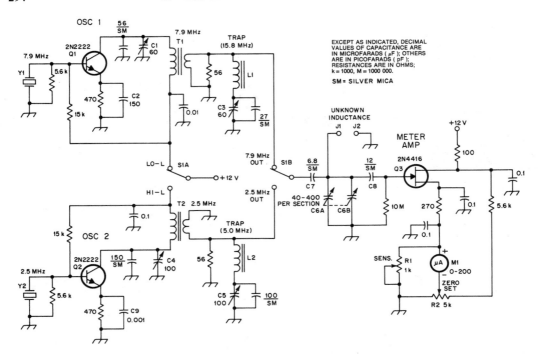

Inductance Meter. Two oscillators, one operating on 2.5 MHz and the second on 7.9 MHz, provide tuning ranges from 10–100 μH and 1–10 μH. C2 and C9 are critical values for establishing the proper amount of oscillator feedback. R1, the 1-K potentiometer, is used to set the meter sensitivity and R2 is a 5-K or 10-K potentiometer used to set the meter to zero when no inductance is connected to J1 and J2. L1 is 24 turns of #26 wire on a T37-6 toroid core and L2 is 40 turns of #30 wire on a T37-2 toroid core. T1 has a primary of 31 turns of #26 wire on a T50-2 toroid core and a secondary of seven turns of #26 wire. T2 has a primary of 19 turns of #26 wire on a FT37-61 form and a secondary of four turns of #26. Crystals Y1 and Y2 are fundamental types with a 30-pF load capacitance. Calibration is done with inductances of known values.—D. DeMaw, "A Tester for Coil Inductance," *QST*, April 1986, pp. 20–22.

L-Pad Power Sampler. Samples output of a transmitter so its waveform can be displayed on an oscilloscope or similar instrument. Circuit is designed for use from 1.8–30 MHz at up to 1000 W of power. Resistor sets are spaced ⅛-in. apart, and BNC connector is at least 3 in. away from the SO239 connectors. The "fuse" is a thin strand of copper, 0.5–0.75 in. long, taken from a "zip cord." Test the circuit at full transmitter power before connecting to oscilloscope or other device.—F. Perkins, "Build the Dixie Whistler," 73 *Magazine*, April 1985, pp. 36–44.

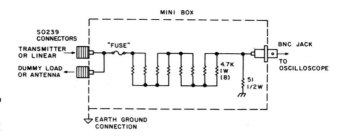

"Nuller" Bridge. Allows tuning up an antenna matching unit for maximum performance without significant radiation from the transmitter. To use, the switch is set to the "TUNE" position and potentiometer RK (50 K) is adjusted until the meter reads zero, at which point the antenna matching unit is tuned for a 50-ohm impedance. D1 is a 1N34A or equivalent.—W. Vissers, "Wheatstones are not Crackers," 73 *Magazine*, September 1985, pp. 50–52.

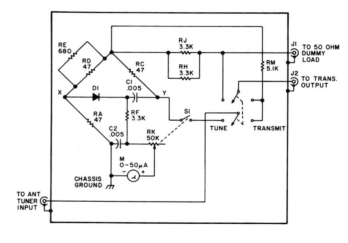

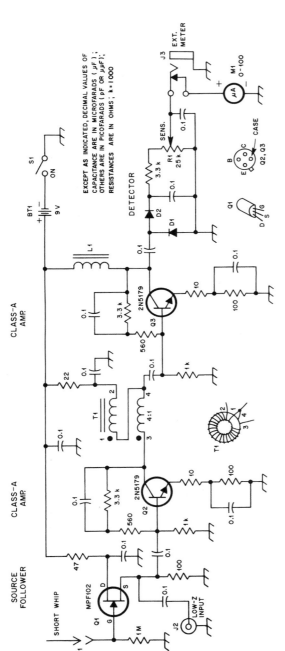

EXCEPT AS INDICATED, DECIMAL VALUES OF CAPACITANCE ARE IN MICROFARADS (μF); OTHERS ARE IN PICOFARADS (pF OR μμF); RESISTANCES ARE IN OHMS; k = 1000

1.8–30-MHz Field Strength Meter. Useful for determining the performance of antennas and transmitting devices and adjusting them properly. The MPF102 JFET at the circuit's input is in a source-follower circuit whose main purpose is impedance transformation to match the "whip" antenna to RF amplifier transistor Q2. The whip antenna can be any general purpose shortened antenna such as that

used by portable radios, walkie-talkies, etc. No tuned circuits are used in this design. T1 is a bifilar-wound transformer with a 4:1 impedance ratio formed from 15 bifilar turns (wires have eight twists per in. before winding) of #26 wire on a FT50-43 toroid core.—D. DeMaw, "Learning to Use Field Strength Meters," *QST*, March 1985, pp. 26–30.

Antenna Test Unit. Unit combines a low power transmitter and SWR meter in a single package to permit "in the field" measurements and adjustment of transmitting antennas intended for operation in the 1800–2000 kHz, 3500–4000 kHz, and 7000–7300 kHz amateur radio bands; coverage can be expanded to other frequency ranges between 1.8–30 MHz. L1 is 20 turns of #24 on a FT-50-61 ferrite core, L2 is 35 turns of #26 on a T68-2 powdered-

iron core, L3 is 21 turns of #24 on a T68-6 core, L4 is 20 turns of #24 on a T50-2 core, L5 is 10 turns of #24 on a T50-2 core, L6 is 25 turns of #26 wire on a T50-2 core, L7 is 13 turns of #24 on a T50-2 core, L8 is 35 turns of #26 on a T50-2 core, and L9 is 17 turns of #24 on a T50-2 core. T1 has a primary of 15 turns of #26 wire on a FT37-43 ferrite toroid with a secondary of four turns.—D. DeMaw, "Field Tester for Antennas," *QST*, March 1986, pp. 40–42.

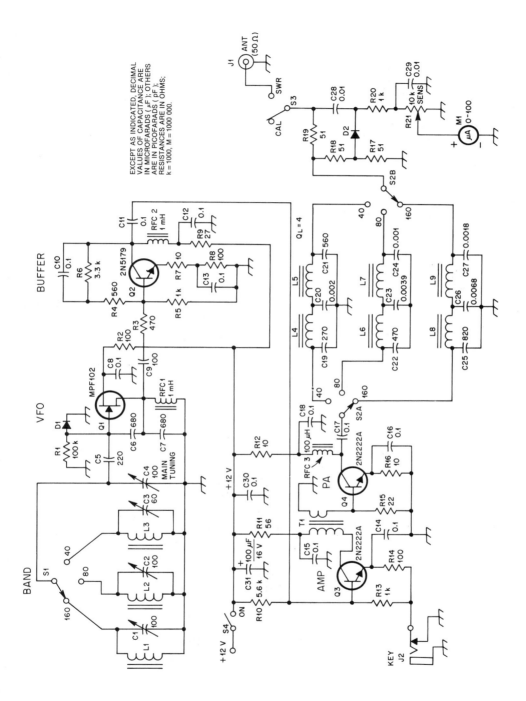

EXCEPT AS INDICATED, DECIMAL
VALUES OF CAPACITANCE ARE
IN MICROFARADS (µF); OTHERS
ARE IN PICOFARADS (pF);
RESISTANCES ARE IN OHMS;
k = 1000, M = 1000 000.

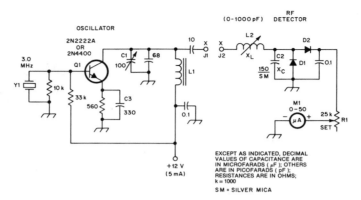

1–1000-pF Capacitance Meter. Resistor R1 in series with the meter helps assure meter linearity. Y1 may also be a 2 MHz or 2.5 MHz crystal; a computer type is recommended. In operation, C3 has a reactance of approximately 155 ohms and is part of a feedback network; the remaining capacitance is within Q1. C2 forms a capacitive divider for the capacitor under test and helps assure a proper meter reading. Diodes D1 and D2 are 1N914 or equivalent. L1 is 18 turns of #28 wire on a FT-37-61 ferrite core. L2 is 21 turns of #28 wire on a 0.25-in. slug-tuned form. Meter is calibrated with capacitors of known values. —D. DeMaw, "Measuring Small Value Capacitors," *QST*, September 1986, pp. 31–33.

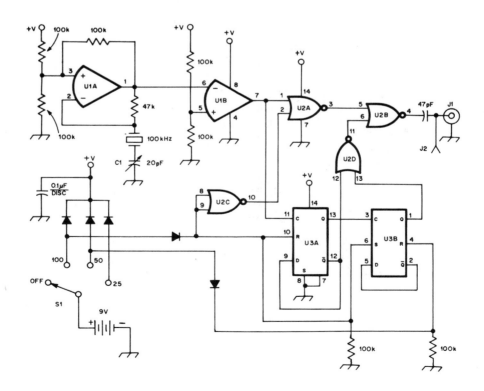

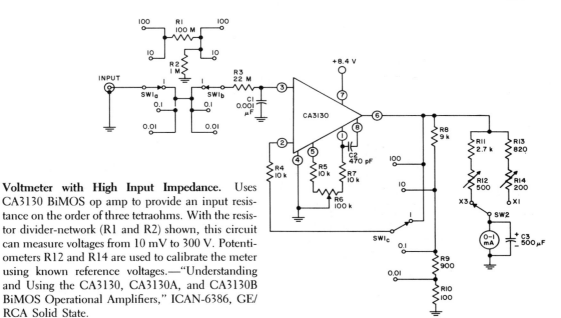

Voltmeter with High Input Impedance. Uses CA3130 BiMOS op amp to provide an input resistance on the order of three tetraohms. With the resistor divider-network (R1 and R2) shown, this circuit can measure voltages from 10 mV to 300 V. Potentiometers R12 and R14 are used to calibrate the meter using known reference voltages.—"Understanding and Using the CA3130, CA3130A, and CA3130B BiMOS Operational Amplifiers," ICAN-6386, GE/RCA Solid State.

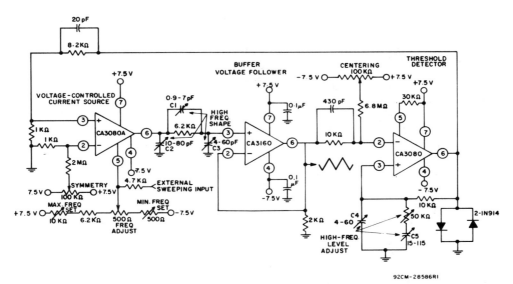

1,000,000/1 Single-Control-Function Generator. Output can range from 1 Hz to 1 MHz. Provision is made for an external sweeping input for display of the output on an oscilloscope or similar instrument.—"CA3080 Operational Transconductance Amplifiers," File #475, GE/RCA Solid State.

◀ **Frequency Standard.** Produces frequency markers at intervals of 25, 50, and 100 kHz at levels strong enough to be useful past 30 MHz. U1 is a LF353N, U2 is a 4001, and U3 is a 4013. All diodes are 1N914 or equivalent. C1 is a 20–30-pF ceramic trimmer capacitor which can be used to adjust the output of the circuit to zero beat with stations WWV and CHU for maximum accuracy.—D. DeMaw, "WWV and CHU in Your Workshop," *QST*, October 1986, pp. 42–44.

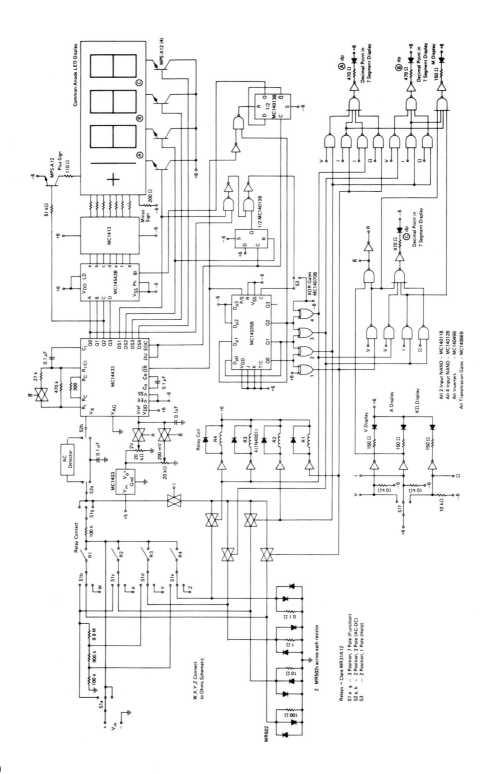

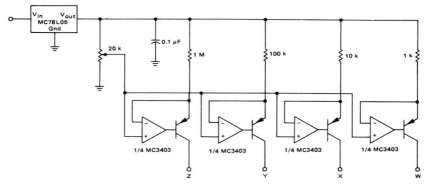

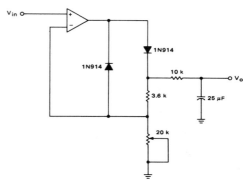

Autoranging Digital Multimeter. MC14433 CMOS analog-to-digital converter device is heart of multimeter capable of measuring AC and DC voltages from 200 mV to 200 V. Circuit can also measure currents from 2 mA to 2 A and resistances from 2 K to 2 M. Circuit has high input impedance. The details of the ohms measurement and AC detection circuitry are shown separately (*above and left*). Power supply requirements are ± 6 V.—"Autoranging Digital Multimeter Using the MC14433 CMOS A/D Converter," AN-769, Motorola, Inc.

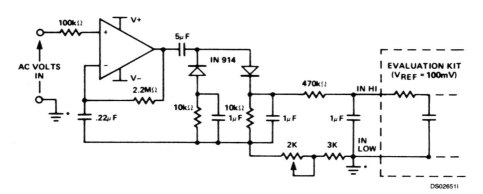

AC-to-DC Converter for Voltmeters.   Designed for voltmeters which can only measure DC voltages, circuit produces a DC voltage corresponding to the average value of an applied AC voltage. Circuit has high input impedance (over 10 M), a bandwidth of 20 Hz to over 5 kHz, and introduces no DC errors into the measurement. Op amp used can be a CA3140 or equivalent.—"Digital Panel Meter Experiments for the Hobbyist," A059, GE/Intersil.

261

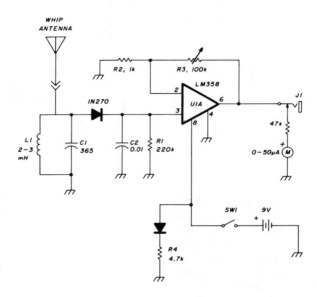

1–30 MHz Field Strength Meter. Effective tuning range depends upon value of L1. A value of 50 μH will cover 1–4 MHz, 3 μH covers 5–16 MHz, and 0.9 μH covers 9–30 MHz. A value of 2.5 mH is suitable for broadband coverage with reduced gain.—S. DeFrancesco, "A Very Sensitive LF or HF Field Strength Meter," *ham radio*, September 1986, pp. 67–68.

26
Transmitting Circuits

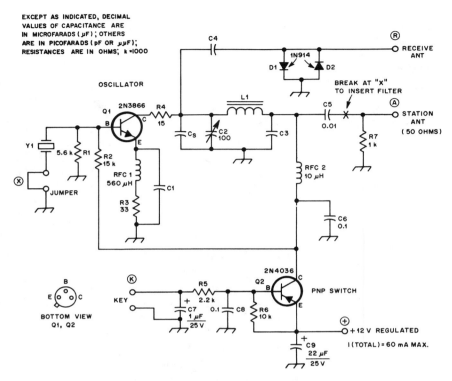

EXCEPT AS INDICATED, DECIMAL
VALUES OF CAPACITANCE ARE
IN MICROFARADS (μF) ; OTHERS
ARE IN PICOFARADS (pF OR μμF);
RESISTANCES ARE IN OHMS; k =1000

Values of L and C for the One-Stage QRP Transmitter (Fig. 4)

Band	C1 (pF)	Cs (pF)	C2 (pF)	C3 (pF)	C4 (pF)	L1 (μH)
3.5 MHz	1000	330	100	1200	100	4.2, 29 turns no. 26 wire on T50-2 toroid core
7.0 MHz	560	180	100	600	47	2.12, 20 turns no. 26 wire on T50-2 toroid core
10.1 MHz	470	100	100	430	33	1.48, 19 turns no. 26 wire on T50-6 toroid core
14.0 MHz	470	68	100	300	27	1.0, 13 turns no. 26 wire on T50-6 toroid core

Note: C2 is a Mouser no. 24AA034 (page 76) 10-mm-diameter trimmer, 15-100 pF. C2, C3 and Cs should be polystyrene or silver mica. All others are disc ceramic.

Four-Band, 250-mW, CW Transmitter. Capable of operation in the 3.5-, 7-, 10.1-, and 14-MHz amateur radio bands depending upon the values of C1, C2, C3, C4, L1, and Cs (as shown in accompanying chart). Circuit can be used for full "break-in" operation without an antenna relay by connecting the receiver-antenna input line to the point labeled "R" in the diagram. Transistor Q2 (2N4036) is a DC switch serving as the keying transistor, R5, C7, and C8 form a shaping network for the CW waveform to prevent "clicks" in the output signal. R7 provides a light load for C5 under all conditions for extra oscillator stability. A VXO can be added at the "jumper" condition indicated by "X" in the left portion of the circuit. Crystal Y1 should be an International Crystal type GP with 20-pF load capacitance.—D. DeMaw, "Simple QRP Gear Versus Good Performance," QST, January 1985, pp. 22–26.

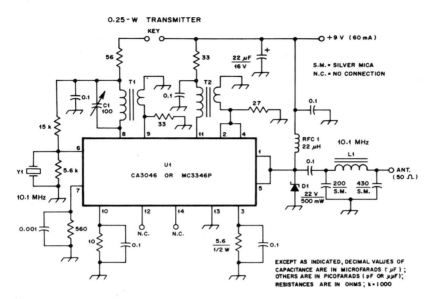

0.25-W TRANSMITTER

S.M. = SILVER MICA
N.C. = NO CONNECTION

EXCEPT AS INDICATED, DECIMAL VALUES OF
CAPACITANCE ARE IN MICROFARADS (μF) ;
OTHERS ARE IN PICOFARADS (pF OR μμF);
RESISTANCES ARE IN OHMS ; k = 1000

10.1-MHz, 250 mW, CW Transmitter. Circuit uses a CA3046 transistor array IC instead of discrete transistors. Capacitor C1 is adjusted for the best sound of the CW note output. D1 is a 1N5251 or equivalent. L1 is formed from 19 turns of #26 wire on a T50-6 toroid core. T1 has a primary of 24 turns of #26 wire on a T50-2 toroid core with a secondary of four turns of #26 wire over the primary. T2 has a primary of 15 turns of #26 wire on a FT50-43 ferrite toroid with a secondary winding of two turns of #26 wire. Crystal Y1 is a fundamental mode crystal cut for the 10.1 MHz amateur radio band.—D. DeMaw, "A Utility IC—The CA3046," *QST*, August 1985, pp. 21–24.

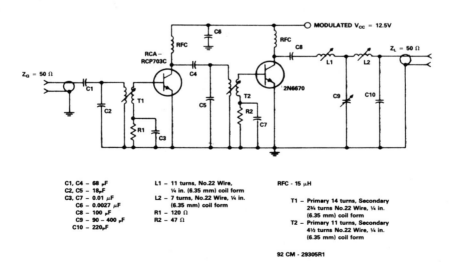

C1, C4 – 68 pF	L1 – 11 turns, No.22 Wire, ¼ in. (6.35 mm) coil form
C2, C5 – 18 pF	L2 – 7 turns, No.22 Wire, ¼ in. (6.35 mm) coil form
C3, C7 – 0.01 μF	
C6 – 0.0027 μF	R1 – 120 Ω
C8 – 100 pF	R2 – 47 Ω
C9 – 90 – 400 pF	
C10 – 220 pF	

RFC - 15 μH

T1 – Primary 14 turns, Secondary 2¾ turns No.22 Wire, ¼ in. (6.35 mm) coil form

T2 – Primary 11 turns, Secondary 4½ turns No.22 Wire, ¼ in. (6.35 mm) coil form

92 CM - 29305R1

27-MHz Transmitter Chain. Provides 4 W of unmodulated, carrier-output power from a 12.5-V supply. The driver and output stages operate class C. Modulation is accomplished by varying the supply voltage. L1 and L2 suppress harmonics of the 27-MHz signal.—"6 and 12 Volt 4 W Transmitters for Class D Citizens Band Radio-Telephony Using the RCA 2N6670," AN-6683, GE/RCA Solid State.

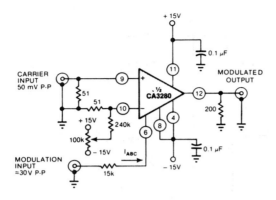

Amplitude Modulator. Circuit uses half of a CA3280 dual op amp. As shown, circuit uses a carrier frequency of 3 MHz and a modulating signal frequency of 10 kHz.—"Dual Variable Op Amp IC, the CA3280, Simplifies Complex Analog Designs," ICAN-6818, GE/RCA Solid State.

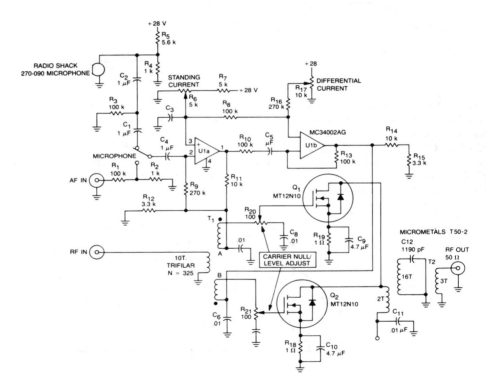

DSB Modulator. MT12N10 TMOS power FETs offer high gain at both AF and RF frequencies to produce a double sideband, suppressed carrier output. RF input is applied to the primary of T1, which applies equal amplitude, opposite phase RF drive for Q1 and Q2. With no AF modulation at points A and B, the opposite phase RF signals cancel each other and no output appears at the 50 ohm output connector. When AF modulation is applied to points A and B, a modulated RF output is obtained. DC stability and low frequency gain are improved by source resistors R18 and R19. A phase inverter using the MC34002AG dual op amp produces the out-of-phase, equal amplitude, AF modulation signals.—L. Lockwood, "Double Sideband, Suppressed Carrier RF Modulator," TMOS *Power FET Design Ideas*, p. 45, Motorola, Inc.

27
VHF/UHF Circuits

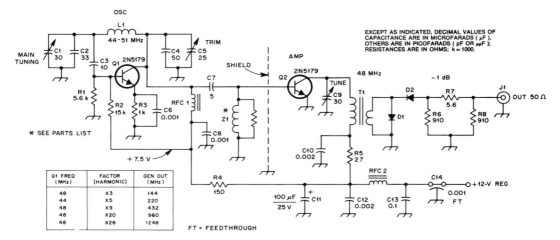

Q1 FREQ (MHz)	FACTOR (HARMONIC)	GEN OUT (MHz)
48	X3	144
44	X5	220
48	X9	432
48	X20	960
48	X26	1248

FT = FEEDTHROUGH

50–1296-MHz RF Signal Generator. Tunable oscillator covering 44–51 MHz produces output harmonics covering the 144-, 220-, 434-, 902-, and 1215-MHz amateur radio bands. The 2N5179 transistor, Q1, is the heart of the LC oscillator covering 44–51 MHz. C1, a 30-pF air variable capacitor, is the main tuning control. C5, a 25-pF ceramic trimmer capacitor, is used to calibrate the oscillator. Q2, another 2N5179, serves as a class C amplifier; class C was used to deliberately enhance the harmonic output. Most other small signal UHF NPN transistors may be substituted. A shield divider is placed between Q1 and Q2 on the component side of the circuit board to prevent unwanted coupling between the circuit sections. D1 and D2 are 1N914 or equivalents. L1 is an air-wound coil of seven turns of #20 wire wound ⅜ in. in diameter and ⅜ in. long. T1 has a primary of 16 turns of #24 wire on a T37-6 toroid form with a secondary of six turns of #24. Z1 is a 33-ohm, 0.5-W carbon resistor wound full (close-wound, single layer) of #30 wire. RFC1 and RFC2 are miniature 10-μH chokes.—D. DeMaw, "Construct a VHF/UHF Signal Generator," *QST*, February 1986, pp. 33–35.

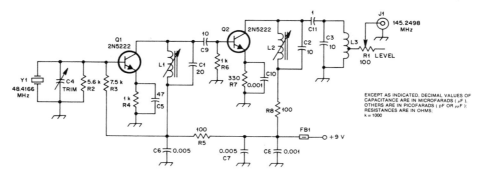

144–148-MHz Oscillator. Output of oscillator, whose frequency is controlled by crystal Y1, is tripled in frequency to produce output in the 144–148 MHz amateur radio band. In circuit shown, Y1 is a third-overtone crystal with a frequency of 48.4166 MHz; the tripled output is 145.25. L1 and C1 are resonant at 48.4166 MHz and the L2/C2 combination is tuned to the third-overtone frequency. L3 is used primarily for impedance matching. C4 is 3–20-pF ceramic trimmer capacitor. L1 is 1.08 μH and L2 is 0.12 μH, both slug-tuned. L3 is five turns of #20 wire on a 0.25-in. coil form, tapped two turns from ground. FB1 is a ferrite bead. SK3246 transistors may be substituted for the 2N5222 units specified.—G. Bonaguide, "CATVI Field-Strength Measurements Made Easy," *QST*, February 1986, pp. 42–46.

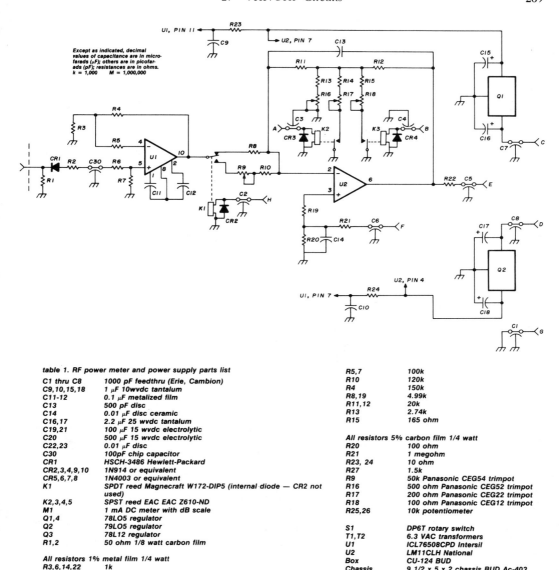

Except as indicated, decimal
values of capacitance are in microfarads (μF); others are in picofarads (pF); resistances are in ohms.
k = 1,000   M = 1,000,000

table 1. RF power meter and power supply parts list			
C1 thru C8	1000 pF feedthru (Erie, Cambion)	R5,7	100k
C9,10,15,18	1 μF 10wvdc tantalum	R10	120k
C11-12	0.1 μF metalized film	R4	150k
C13	500 pF disc	R8,19	4.99k
C14	0.01 μF disc ceramic	R11,12	20k
C16,17	2.2 μF 25 wvdc tantalum	R13	2.74k
C19,21	100 μF 15 wvdc electrolytic	R15	165 ohm
C20	500 μF 15 wvdc electrolytic		
C22,23	0.01 μF disc	All resistors 5% carbon film 1/4 watt	
C30	100pF chip capacitor	R20	100 ohm
CR1	HSCH-3486 Hewlett-Packard	R21	1 megohm
CR2,3,4,9,10	1N914 or equivalent	R23, 24	10 ohm
CR5,6,7,8	1N4003 or equivalent	R27	1.5k
K1	SPDT reed Magnecraft W172-DIP5 (internal diode — CR2 not used)	R9	50k Panasonic CEG54 trimpot
		R16	500 ohm Panasonic CEG52 trimpot
K2,3,4,5	SPST reed EAC EAC Z610-ND	R17	200 ohm Panasonic CEG22 trimpot
M1	1 mA DC meter with dB scale	R18	100 ohm Panasonic CEG12 trimpot
Q1,4	78L05 regulator	R25,26	10k potentiometer
Q2	79L05 regulator		
Q3	78L12 regulator	S1	DP6T rotary switch
R1,2	50 ohm 1/8 watt carbon film	T1,T2	6.3 VAC transformers
		U1	ICL76508CPD Intersil
All resistors 1% metal film 1/4 watt		U2	LM11CLH National
R3,6,14,22	1k	Box	CU-124 BUD
		Chassis	9 1/2 x 5 x 2 chassis BUD Ac-403

UHF RF Power Meter. Offers a 30-dB range from −15 dBm to −45 dBm with good temperature stability. Useful frequency coverage ranges from about 2 MHz well into the GHz area, with the circuit designed primarily for work in the UHF region. Heart of the unit is a HSCH-3486 zero-bias Schottky diode (CR1) which is used as the detector. Its response curve is logarithmic from −50 dBm to −20 dBm; above −20 dBm it becomes increasingly nonlinear. This is followed by a pair of operational amplifiers, an ICL7650 (U1) and a LM11 (U2), to amplify the resulting signal. Original article gives details on device calibration and use. Accompanying table gives values for parts in the schematic.—R. Six, "Wide Range RF Power Meter," *ham radio*, April 1986, pp. 24–28.

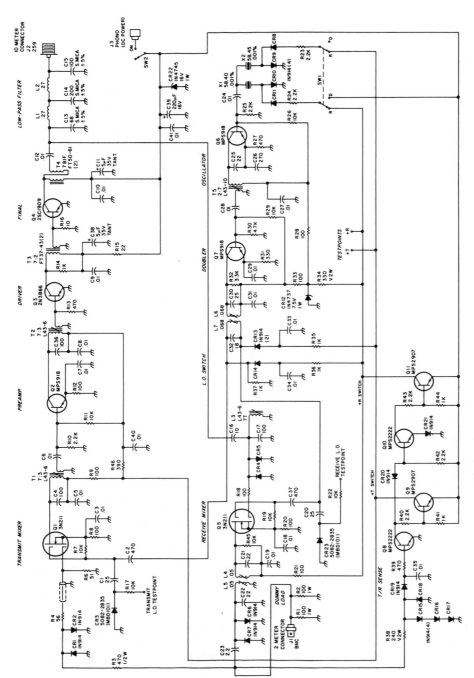

144–29-MHz Transverter. Allows an amateur radio FM transceiver operating near 146 MHz to cover the 29-MHz FM segment of the 28–29.7-MHz amateur radio band. One W of input at 146 MHz will produce approximately 3.5 W at 29 MHz. A small sample of the 146-MHz RF energy is mixed with the transverter's local oscillator (approximately 116.80 MHz) in the transmit mixer to produce a difference frequency of 29 MHz. Tuned circuits reject unwanted mixer-outputs before final amplification. On receive, the process is reversed; the 29-MHz input signal is mixed to produce at 146-MHz output. Accompanying chart gives coil and inductor data.—F. Perkins, "Two to Ten," *73 Magazine*, June 1986, pp. 38–45.

Coil Reference #	Nominal Inductance	# of Turns-N	Wire Gauge AWG.	Inside Coil Diameter D	Coil Length L	Spacing S	Mandrel
L1	.27 uH	6	#18 (1)	.45" (11.4)	.45" (11.4)	N/A	½-13
L2	.27 uH	6	#18 (1)	.45" (11.4)	.45" (11.4)	N/A	½-13
L4	+ .05 uH	3	#18 (1)	.22" (5.6)	.15" (3.8)	.2" (5.1):L5	¼-20
L5	− .05 uH	3	#18 (1)	.22" (5.6)	.25" (6.35)	see above	¼-20
L6	+ .068 uH	3	#18 (1)	.28" (7.1)	.20" (5.1)	.2" (5.1):L7	⅜-16
L7	− .068 uH	3	#18 (1)	.28" (7.1)	.25" (6.35)	see above	⅜-16

Parenthetical values are in millimeters.

L3   7 turns of #30 (.25) close-wound at the base of a Micromet-
        als, Inc. L43-6 form, .40 uH.

T1,T2 Primary—7 turns of #30 (.25) close-wound at the base of a
        Micrometals, Inc. L43-6 form, .40uH; secondary—3 turns of
        #30 (.25) wound over primary.

T3   Primary—7 turns of #26 (.40) wound 75% around two
        stackedFT37-43 ferrite toroid cores; secondary—2 turns of
        #26 (.40) wound over the middle of the primary.

T4   7 bifilar turns of #26 (.40) wound 75% around two stacked
        FT50-61 ferrite toroid cores; interconnect for 1:4 step-up.

T5   Primary—7 turns of #30 (.25) close-wound at the base of a
        Micrometals Inc., L43-10 form, .34 uH; secondary—2 turns
        of #30 (.25) wound over primary.

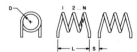

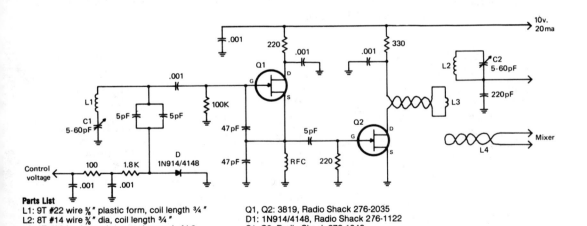

Parts List

L1: 9T #22 wire ⅜" plastic form, coil length ¾"
L2: 8T #14 wire ⅜" dia, coil length ¾"
L3: 2T #22 hook-up wire wound on end of L2
L4: 1T #14 wire loosely coupled to L2 (~ ½" from L2)
ch: 1 watt high value R single layer #30 wire

Q1, Q2: 3819, Radio Shack 276-2035
D1: 1N914/4148, Radio Shack 276-1122
C1, C2: Radio Shack 272-1340
All fixed capacitors are disc ceramics
All resistors are ¼ watt

50-MHz VFO. Frequency-shift keying is accomplished by varying conduction of a 1N914 diode in
series with a 10-pF capacitor paralleling the VFO
frequency control LC. The variable reactance produces a frequency shift of 900 kHz. Output is induc-
tively coupled to the transmitter mixer-stage as indicated.—J. Reed, "A Simple Phase-Locked-Loop
Method," *CQ*, June 1985, pp. 26–29.

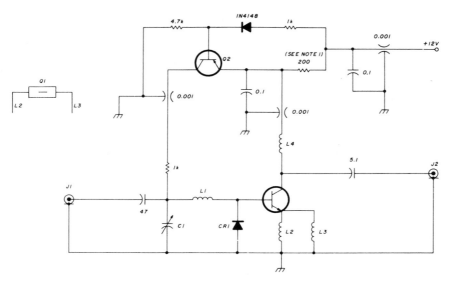

902–928 MHz Receiving Preamplifier. Provides approximately 12 dB of gain with a noise figure of 2.5–3 dB for reception of signals in the newly authorized 902–928 MHz amateur radio band. C1 is a 1–10-pF air variable capacitor. CR1 is a HP 5082-2810 hot carrier diode or equivalent. L1 is the base lead of Q1 plus ⅛ in. of #24 wire to C1. L2 and L3 are the full leads of Q1 (MRF901 or equivalent; this is the unlabeled transistor in the schematic) tacked end to end as shown. L4 is two turns of #24 wound 0.10 in. (inside diameter). Q2 is a 2N2907 or equivalent high gain PNP transistor.—J. Reisert, "VHF/UHF World," *ham radio*, April 1986, pp. 83–92.

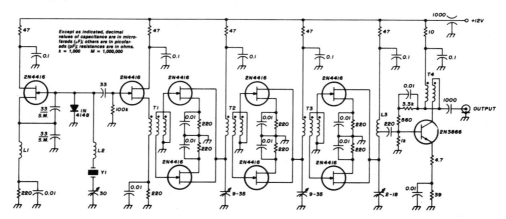

Local Oscillator for 144-MHz SSB Transmitter. Circuit is a VXO running at a nominal frequency of 18,025 kHz followed by three doubler stages and a buffer amplifier. Output in the intended 144.150–144.300 MHz coverage is about +17 dBm. Y1 operates at 18,025 kHz, parallel resonant, with 20-pF load capacitance. L1 is 33 turns of #24 wire on a FT50-61 form. L2 is 35 turns of #24 on a T50-6 form. L3 is four turns of #18 wound 0.25 in. in diameter and 0.5 in. long tapped one-half turn from the "cold" end. T1 is 12 turns of #30 wire, trifilar wound, on a T37-6 core. T2 is 11 turns of #30, trifilar wound, on a T37-6 form. T3 is seven turns of #30, trifilar wound, on a T37-6 form. T4 is five turns of #24 wire, bifilar wound, on a FT37-61 form.—N. Bernstein, "Two Meter Transmitter Uses Weaver Modulation," *ham radio*, July 1985, pp. 12–19.

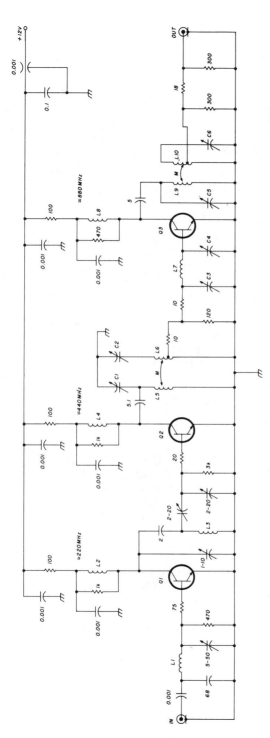

Frequency Multiplier with 800–900 MHz Output.
Three doubler circuits take an input of over 100 MHz to produce an output in the 800–900 MHz range. Output is approximately 10–20 mW for an input of 5–10 mW. Circuit output is "clean" of spurious emissions and circuit alignment is relatively easy. Circuit should be entirely enclosed in a shielded box. C1 through C6 are all 1–10-pF air variable capacitors. L1 is 12 turns of #24 wire wound 0.10 in. in diameter. L2 and L3 are both four turns

of #24 wound 0.25 in. long and 0.10 in. in diameter. L4 is a 0.1-μH RF choke. L7 is four turns #24 wire wound 0.10 in. in diameter and 0.25 in. long. L8 is eight turns of #24 wound 0.10 in. in diameter and 0.25 in. long. L9 and L10 are the same as L5 and L6 with L10 tapped 0.25 in. up. Q1 and Q2 are either NE73432B or 2N5179 while Q3 is a NE02132.—J. Reisert, "VHF/UHF World," *ham radio*, April 1986, pp. 83–92.

273

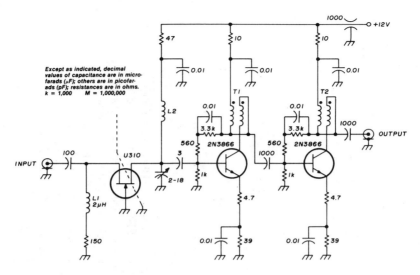

Except as indicated, decimal values of capacitance are in microfarads (μF); others are in picofarads (pF); resistances are in ohms. k = 1,000   M = 1,000,000

Post-Mixer Amplifier for VHF Transmitter.
Provides 30-dB gain to the output of the mixer stage of a 144–148-MHz SSB transmitter. Designed to operate with an input signal of − 10 dBm and produces an output of 100 mW. Amplifier is a three-stage device with a grounded gate FET followed by two broadband, bipolar class A stages. L1 is 34 turns of #26 wire closely wound on a 100-K, 0.5-W resistor.

L2 is four turns of #20 wire wound 0.25 in. in diameter and 0.5 in. long. T1 and T2 are five turns of #24 wire bifilar wound on a FT37-61 core. The 2N3866 transistors should be fitted with heat sinks.—N. Bernstein, "Two Meter Transmitter Uses Weaver Modulation," *ham radio*, July 1985, pp. 12–19.

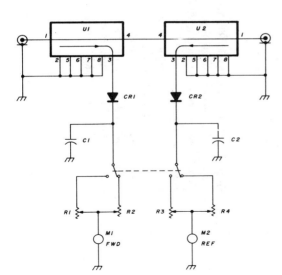

50–100 MHz Dual Wattmeter. U1 and U2 are 11.5-dB directional couplers (Mini-Circuits PDC-10-1 or equivalent) whose sampled port is − 11.5 dB from the line; both are encapsulated in miniature metal cans. Their coupling ratio is flat to within ±0.6 bD from 500 kHz to 500 MHz, and maximum power on the throughline is 3 W from 5–500 MHz. The remainder of the circuit is a typical RF voltmeter. C1 and C2 are 0.01-μF ceramic capacitors. CR1 and CR2 are HP-2800 or equivalent Schottky diodes. M1 and M2 are 0–100-μA meters. R1 through R4 are 50-K trimmer potentiometers.—B. Lombardi, "A 50–500 MHz Dual Wattmeter," *ham radio*, July 1985, pp. 67–70.

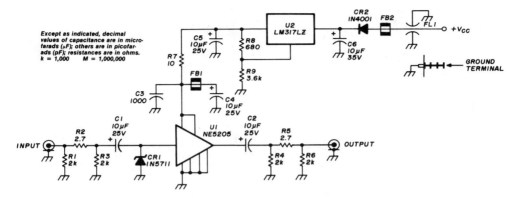

Wideband VHF Small Signal RF Amplifier. Based upon Signetics NE5205 RF amplifier IC, circuit offers 20 dB of gain at approximately 600 MHz with usable gain to 1.2 GHz. Wideband noise figure is about 5 dB. Output power at the 1-dB compression point is +6 dBm. Total supply current is about 35 mA and the minimum supply voltage is +10 V. Circuit must be housed in a shielded enclosure. FB1 and FB1 are ferrite beads, type FB-43-101 or equivalent. Original article includes construction details and PC board artwork.—M. Gruchalla, "NE5205 Wideband RF Amplifier," *ham radio*, September 1986, pp. 30–38.

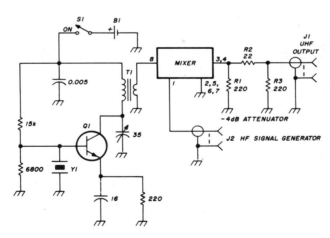

Harmonic Mixer with Output to 500 MHz. Heart of circuit is Mini-Circuits Lab's SBL-1 (or equivalent) diode-ring mixer (labeled "mixer" in schematic). When driven at high signal levels, these produce harmonics well into the VHF range. Input signal from a HF signal generator causes generation of VHF signals. Supply voltage by B1 should be 9V. Q1 is a 2N918 or equivalent. T1 has a primary of 12 turns of #26 wire and a secondary of two turns of #26. Y1 is a third-overtone 50-MHz crystal.—P. Bertini, "Harmonic Mixer for VHF Signal Generation," *ham radio*, March 1985, pp. 40–41.

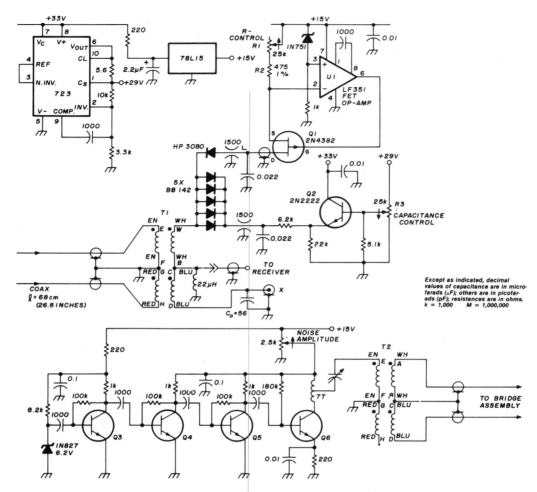

VHF Noise Bridge. Allows measuring resistances and reactances in antenna systems and transmission lines to within 3 ohms at 146 MHz. T1 and T2 must have close magnetic coupling between primary and secondary; it consists of four tightly twisted #24 wires of 0.5 mm diameter. Two opposite wires of the bundle are primary and secondary windings. The bundle is threaded about 3.5 times through a T50-10 core. Original article gives details on alignment and construction.—A. Popodi, "A VHF Noise Bridge," *ham radio,* July 1986, pp. 10–19.

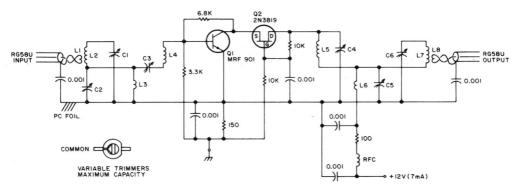

146-MHz Receiving Preamplifier. Delivers 20 dB of gain at 146 MHz; circuit was originally designed for reception of "downlink" signals from amateur radio satellites operating near 146 MHz. The hybrid cascade design takes advantage of the gain and low noise figures of the MRF901, while the 2N3819 improves the stability and overload resistance of the circuit. Inductors L2, L4, L5, and L7 are all five turns of #14 wire wound ⅜ in. in diameter and 7/16 in. long on a ¼ in. diameter rod. L3 and L6 are three turns of #14 wire wound 5/16 in. in diameter and ¼ in. long on a 3/16 in. diameter rod. L1 is two turns of #24 plastic-insulated wire formed over the end of L2, and L1 is two turns of #24 plastic-insulated wire formed over the end of L7. RFC is approximately 30 turns #30 wire on a 0.5-W, 1-K resistor wound in two layers.—J. Reed, "Hear, Hear!," 73 *Magazine*, April 1985, pp. 20–22.

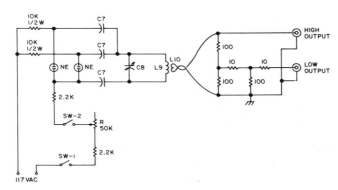

144–148-MHz Noise Source. Uses the noise produced by the discharge of two Ne-2H neon bulbs as a noise source for alignment of receiving systems operating in the 144–148-MHz amateur radio band. The 50-K potentiometer is used to adjust the amplitude of the noise output and the switch S2, a "push-to-break" type, manually activates the bulbs. L9 is six turns of #14 wire wound ⅜ in. in diameter and ½ in. long on a ¼-in. diameter rod. L10 is two turns of #24 plastic-insulated wire formed over the center of L9.—J. Reed, "Hear, Hear!," 73 *Magazine*, April 1985, pp. 20–22.

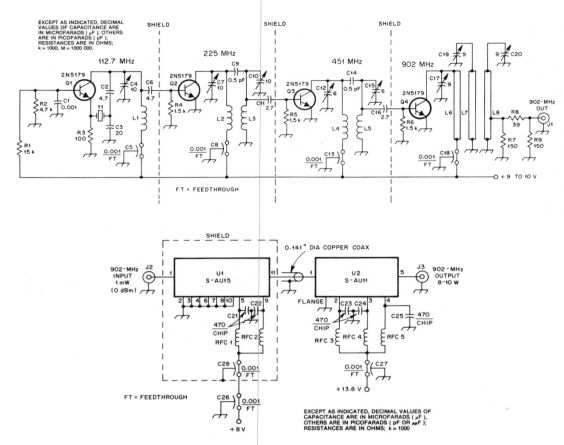

10-W CW Transmitter for 902 MHz. Designed for recently authorized 902-MHz amateur radio band, transmitter has built-in and separate exciter, driver and power amplifier, and power supply and keyer sections. First oscillator is tuned to approximately 112 MHz and three doubler circuits are used to produce a 902-MHz output. Output power of the exciter is only a few milliwatts, and a bandpass filter is used to suppress output on undesired frequencies. The power amplifier section consists of two prepackaged class C "gain blocks" manufactured by Toshiba (S-AU15 and S-AU11). The first requires about 1 mW of input to produce a 100-mW output, and the second requires 100 mW of input to produce a 10-W output. The power supply provides the voltages required by the various stages. Original article gives details on construction techniques and precautions which must be taken at these frequencies. In the exciter section, L1 is five turns of #22 tinned wire, 0.228 in. in diameter (no. 1 drill), spaced one-wire diameter. L2 and L3 are four turns of #18 tinned wire, 0.250 in. in diameter, spaced one-wire diameter. L4 and L5 are two turns #18 tinned wire, 0.250 in. in diameter, spaced one-wire diameter. L6, L7, and L8 are inductors made from copper strap, one in. long by ⅛ in. wide; original article has construction details for these. In the driver and power amplifier sections, RFC1–RFC5 are all three turns of #28 tinned wire, 0.10 in. in diameter, spaced one-wire diameter. In the power supply and keyer section, T1 has a 117-V primary and a 15–24V, 2-A secondary. U5 is a 50-PIV, 4-A bridge-rectifier module.—D. Hilliard, "A CW Transmitter for 902 MHz," *QST*, March 1986, pp. 32–39.

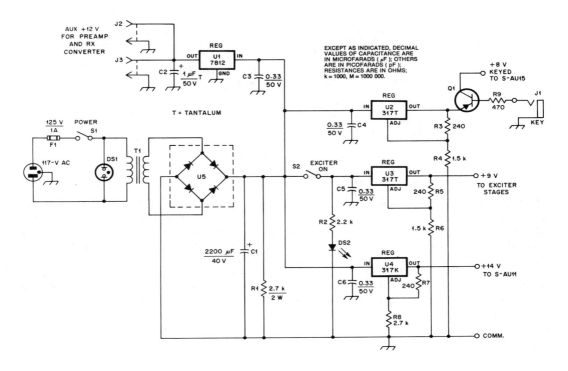

Schematic diagram of the power
supply and keyer for the 902-MHz transmitter. Resistors are ¼ W unless noted.
Capacitors are 50-V ceramic types unless
noted. Capacitors marked with polarity are
electrolytic.

DS1—Neon pilot lamp, 117 V.
DS2—LED.
J1—1/4-in female phone jack, chassis
  mount.
J2, J3—Chassis-mount female phono jack.
S1, S2—SPST toggle switch, 125 V.
Q1—PNP transistor capable of switching
  10 V at 150 mA (see text).
T1—Power transformer; primary, 117-V;
  secondary, 15-24 V, 2 A (see text).
U1—12-V, 1.5-A 3-terminal regulator (LM7812 or equiv).
U2, U3—Adjustable 1.5-A 3-terminal regulator (LM317T or equiv).
U4—Adjustable 3-A 3-terminal regulator (LM317K or equiv).
U5—50-PIV, 4-A bridge rectifier module.

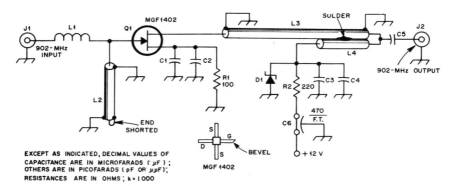

EXCEPT AS INDICATED, DECIMAL VALUES OF CAPACITANCE ARE IN MICROFARADS (µF) ; OTHERS ARE IN PICOFARADS (pF OR µµF); RESISTANCES ARE IN OHMS ; k = 1 000

902-MHz Receiving Preamplifier. Provides approximately 17 dB of gain and a noise figure of 0.75 dB for reception on the recently established 902-MHz amateur radio band. Design is broadband and can be used as is for reception approximately 200 MHz above and below 902 MHz. L1 is four turns of #30 wire, 0.095 in. in diameter (no. 41 drill), spaced one-wire diameter. L2, L3, and L4 are made from 0.141-in. miniature semirigid coaxial cable

such as RG-402 or equivalent; this cable is copper jacketed, has a Teflon dielectric, and a silver-plated, solid center conductor. L2 is a 2.3-in. section, L3 is a 1.6-in. section, and L4 is a 0.465-in. section. A DXL2501 may be substituted for the MGF1402. Original article contains details on component layout and construction.—D. Hilliard, "A 902–144 MHz Receive Converter," October 1985, pp. 21–26.

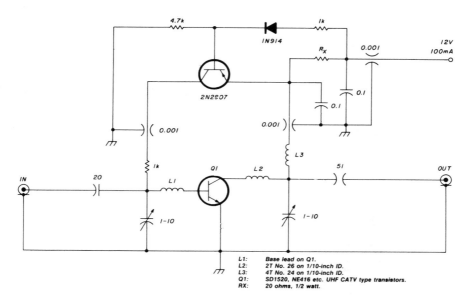

L1:	Base lead on Q1.
L2:	2T No. 26 on 1/10-inch ID.
L3:	4T No. 24 on 1/10-inch ID.
Q1:	SD1520, NE416 etc. UHF CATV type transistors.
RX:	20 ohms, 1/2 watt.

UHF Linear Transmit Amplifier. UHF bipolar transistors designed for CATV applications are used for this circuit intended for use in the 902–928-MHz amateur radio band. Gain is typically 13 dB per stage with a 1-dB compression point of 300 mW. Two stages can be used for added gain and output power.

Q1 is a SD1520, NE416, or equivalent. RX is 20 ohms, 0.5 W. L1 is the base lead of Q1. L2 is two turns of #26 wire wound 0.10-inch in diameter. L3 is four turns of #24 wound 0.10 in. in diameter.—J. Reisert, "VHF/UHF World," *ham radio*, April 1986, pp. 83–92.

28

Voltage Converter and Booster Circuits

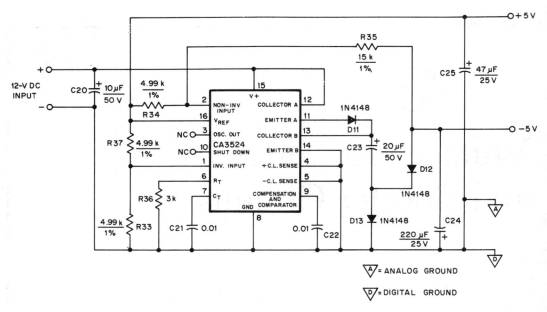

DC-to-DC Converter. CA3524 regulating pulse-width modulator IC provides dual-polarity 5-V output from a 12-V DC input.—R. Arndt and J. Fikes, "SuperSCAF and Son—A Pair of Switched Capacitor Audio Filters," *QST*, April 1986, pp. 13–19.

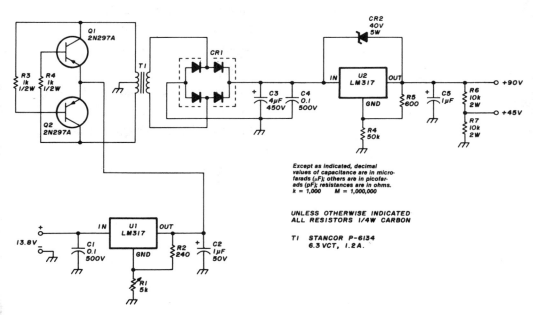

13.8-V to 90-V Power Inverter. Accepts ±13.8-V DC input and delivers +45 and +90 V. With 13.8 V at U1, R1 is adjusted until the output voltage at C3 is +140 V maximum. This occurs when Q1 and Q2 are biased into oscillation by R3 and R4, which produces high voltage AC at the secondary of T1. This AC voltage is rectified by CR1, filtered by C3, and applied to the input of U2. 120-V should be present at the input of U2. T1 is a Stancor P-6134 with a 6.3-V, 1.2-A, center-tapped secondary.—M. Starin, "Get on Six Meters—The Inexpensive Way," *ham radio*, March 1985, pp. 91–95.

+8 and −4 V DC from +5 V DC Source. Intersil ICL7660 is used to provide +8 and −4 V DC from a TTL-standard +5-V DC source. Original purpose of this circuit was to power an op amp in a TTL circuit design.—"CA3310, CA3310A CMOS 10-Bit Analog to Digital Converter with Internal Track and Hold," File # 1851, GE/RCA Solid State.

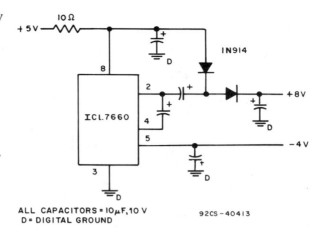

ALL CAPACITORS = 10μF, 10 V
D = DIGITAL GROUND

92CS-40413

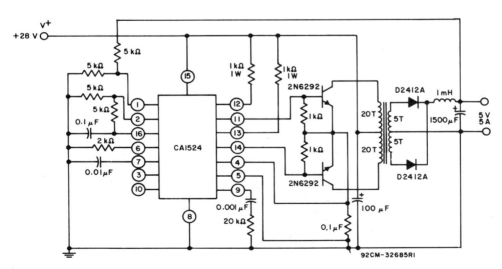

92CM-32685RI

Push-Pull Converter. Produces a 5-V, 5-A output from a 28-V input. The output stages of the CA1524 PWM IC provide the drive for transistors Q1 and Q2 in push-pull application. Since the internal flip-flop divides the oscillator frequency by two, the oscillator must be set at twice the output frequency. Current limiting is done in the primary of transformer T1 so that the pulse width will be reduced if transformer saturation should occur.—"Application of the CA1524 Series Pulse-Width Modulator ICs," ICAN-6915, GE/RCA Solid State.

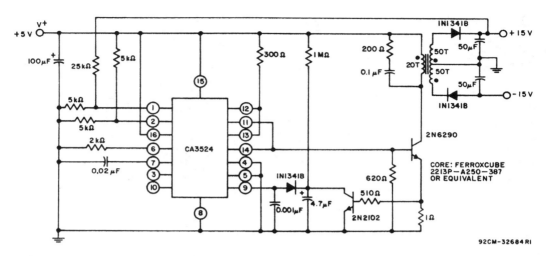

92CM-32684RI

Flyback Converter. Produces a dual polarity 15-V output at 20 mA from a + 5 V regulated source using a CA3524 PWM IC. Reference voltage is provided by the input, and the internal reference generator is unused. Current limiting is accomplished by sensing current in the primary line and resetting the soft-start circuit.—"Application of the CA1524 Series Pulse-Width Modulator ICs," ICAN-6915, GE/RCA Solid State.

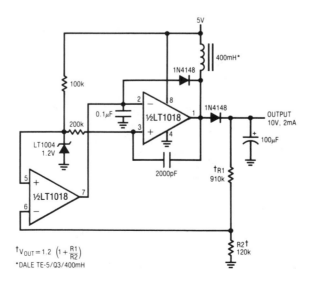

Regulated Up Converter. 5-V input voltage is boosted to a 10-V, 2-mA output by use of both sections of LT1018 comparator IC. Output voltage is determined by multiplying 1.2 by the sum of 1 and the result of R1 divided by R2.—"LT1017/LT1018 Micropower Dual Comparator," data sheet, Linear Technology Corporation.

$$\dagger V_{OUT} = 1.2 \left(1 + \frac{R1}{R2}\right)$$

*DALE TE-5/Q3/400mH

Voltage Doubler. LTC1044 switched capacitor voltage converter boosts 3-V battery supply to 6 V. Efficiency exceeds 90% for load currents below 1.75 mA.—"Power Conditioning Techniques for Batteries," Application Note 8, Linear Technology Corporation.

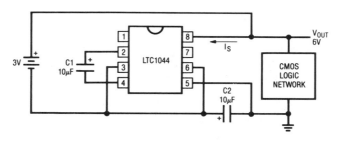

Negative Voltage Generator. LTC1044 switched-capacitor voltage converter device produces a negative output voltage of approximately −8.5 V from a +9-V supply voltage. Output regulation is excellent.—"Power Conditioning Techniques for Batteries," Application Note 8, Linear Technology Corporation.

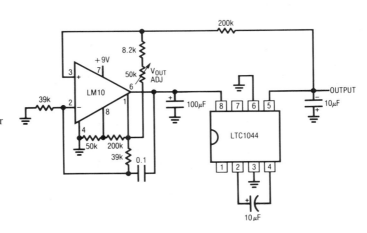

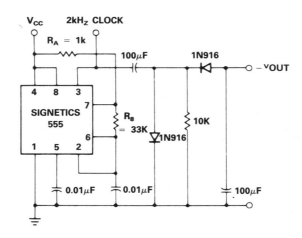

ALL RESISTOR VALUES ARE IN OHMS

Voltage Polarity Inverter. 555 timer IC allows transformerless conversion of one DC voltage to another of the opposite polarity. The negative output voltage tracks the positive input voltage in a linear fashion but about 3 V lower in magnitude.—"NE555 and NE556 Applications," AN170, Signetics Corporation.

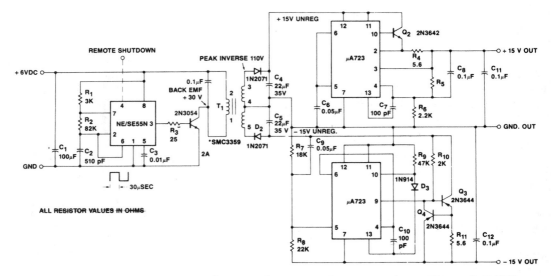

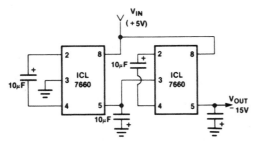

Regulated DC-to-DC Converter. Produces +15 and −15 V DC outputs from +6-V input; +5-V input may be also used. Line and load regulation is 0.1% or better. Transformer T1 is a SMC3359 type from Shafer Magnetics.—"NE555 and NE556 Applications," AN170, Signetics Corporation.

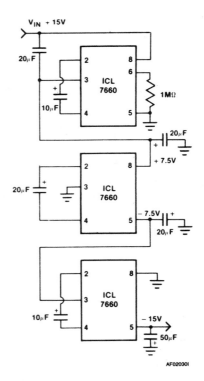

+5 V to −15 V Converter. Two ICL7660 CMOS voltage converters produce a −15-V output from a TTL-level +5-V input.—"Principles and Applications of the ICL7660 CMOS Voltage Converter," A051, GE/Intersil.

+15 V to −15 V Voltage Inverter. Three ▶ ICL7660 CMOS voltage-converter ICs are used to invert a +15 V input into a −15 V output. Intermediate conversion steps to 7.5-V are used, and the output impedance is about 250 ohms.—"Principles and Applications of the ICL7660 CMOS Voltage Converter," A051, GE/Intersil.

AF02030I

29

Voltage Reference Circuits

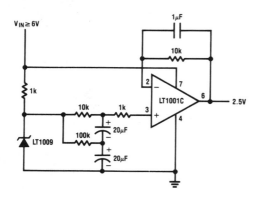

2.5-V Buffered Reference. Input voltage of 6 V or greater produces a stable reference voltage of 2.5 V. Buffered output isolates the LT1009 voltage reference device from following circuit. Output is low in "noise."—"LT1009 Series 2.5 Volt Reference," data sheet, Linear Technology Corporation.

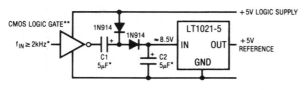

*FOR HIGHER FREQUENCIES C1 AND C2 MAY BE DECREASED.
**PARALLEL GATES FOR HIGHER REFERENCE CURRENT LOADING.

5-V Reference from 5-V Supply Voltage. LT1021-5 voltage reference IC is configured to work from a + 5-V logic supply voltage. Input frequency to the CMOS inverter is in excess of 2 kHz. —"LT1021 Precision Reference," data sheet, Linear Technology Corporation.

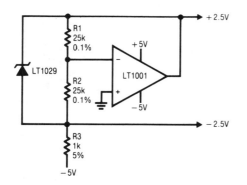

Split 2.5-V Reference. A − 5-V reference voltage is "split" into dual polarity 2.5-V output. LT1029 voltage reference is rated at 5 V.—"LT1029/ LT1029A 5 V Badgap Reference," data sheet, Linear Technology Corporation.

10-V Single Supply Reference. LM329 voltage reference provides a 10-V reference voltage from a + 15-V supply voltage. Output is buffered by the LT1001 operational amplifier.—"LM129/LM329 6.9 V Precision Voltage Reference," data sheet, Linear Technology Corporation.

Index